Developments in Hydrobiology 192

Series editor
K. Martens

Lagoons and Coastal Wetlands in the Global Change Context: Impacts and Management Issues

Lagoons and Coastal Wetlands in the Global Change Context: Impacts and Management Issues

Selected papers of the International Conference "CoastWetChange", Venice, 26–28 April 2004

Edited by

P. Viaroli[1], P. Lasserre[2] and P. Campostrini[3]

[1]*Dipartimento di Scienze Ambientali, University of Parma, Viale Usberti 33 A, 43100 Parma, Italy*
[2]*Pierre and Marie Curie University - Paris VI, France*
[3]*CORILA, Venice, Italy*

Reprinted from Hydrobiologia, Volume 577 (2007)

Springer

Library of Congress Cataloging-in-Publication Data

A C.I.P. Catalogue record for this book is available from the Library of Congress.

ISBN 978-90-481-7501-7
ISBN 978-1-4020-6008-3 (e-book)

Published by Springer,
P.O. Box 17, 3300 AA Dordrecht, The Netherlands

Cite this publication as Hydrobiologia vol. 577 (2007).

Cover illustration: Lagoon of Venice (photo courtesy of Archivio Magistrato alle Acque di Venezia - Consorzio Venezia Nuova)

Printed on acid-free paper

TABLE OF CONTENTS

Hydrobiologia (2007) 577:1–3
DOI 10.1007/s10750-006-0412-9

LAGOONS AND COASTAL WETLANDS

Preface

Pierluigi Viaroli · Pierre Lasserre · Pierpaolo Campostrini

Lagoons and coastal wetlands are among the most common environments in the transitional zone located between terrestrial ecosystems and adjacent seas. Their persistence and ecosystem processes are controlled by complex interactions among stressors and fluxes of material between land, ocean and atmosphere. As a result, coastal zones are among the most changeable and vulnerable environments on Earth. Among other, natural factors that have the largest impact on coastal lagoons and wetlands are sea-level rise, precipitation and river runoff, and storminess (Crossland et al., 2005; Eisenreich, 2005). Natural stressors are interconnected in many ways and are often associated with human impact. In recent decades, most coastal ecosystems have experienced strong anthropogenic pressures, due to progressive human migration from continental areas. At present, nearly 40–50% of the population lives within 100 km of the coastline, including some of the world's largest cities (Crossland et al., 2005).

The impact of human activities on the variability of coastal systems is considerable, and usually leads to deterioration and losses of marine resources, standing stocks and coastal landscape. These pressures have dramatically increased in the last few decades and will continue and evolve, especially in developing countries. Therefore, lagoons and coastal wetlands are expected to be affected by growing modification, i.e. urbanisation, exploitation for aquaculture, marinas and tourism, as well as by large-scale climatic changes.

Most coastal lagoons and their watersheds are influenced by sea eustatism and are subjected to a natural subsidence that has been accelerated by marshland reclamation, groundwater and natural gas extraction. The combination of subsidence and sea-level rise may not be balanced by accretion of coastal wetlands, resulting in increased flooding and saltwater intrusion into freshwater wetlands. Furthermore, a rising sea level combined with more frequent storms and associated surges are likely to cause enhanced coastal erosion. Wetlands and barrier islands reduce storm surges and weaken their energy, to a greater extent than artificial sea defences.

An increased variety of land uses have contributed to increased changes in watershed

Guest editors: P. Viaroli, P. Lasserre & P. Campostrini
Lagoons and Coastal Wetlands in the Global Change Context: Impacts and Management Issues

P. Viaroli (✉)
Dipartimento di Scienze Ambientali, University of Parma, Viale Usberti 33 A, Parma 43100, Italy
e-mail: pierluigi.viaroli@unipr.it

P. Lasserre
Pierre and Marie Curie University, Paris VI, France

P. Campostrini
CORILA, Venice, Italy

structure and hydrographic networks. Overall, these alterations influence coastal wetlands and nearshore coastal waters through spatial dependent and time-lagged processes that control the delivery of nutrients and pollutants (Valiela et al., 1997).

Coastal lagoons and wetlands are recognised as highly unpredictable environments. There is evidence that within certain thresholds, marine communities and ecosystems are resilient to environmental changes and can buffer against external stresses. However, resilience and buffering capacities do not follow linear behaviour, but rather undergo sudden and exponential responses. Therefore, an increasing stress—e.g. by physical and chemical stressors—can result in rapid regime shifts and irreversible deterioration of the aquatic ecosystems. Assessments are further hampered by lack of historical time series, Venice lagoon, Wadden Sea and Chesapeake Bay being probably the only environments where this analysis has been attempted.

Coastal lagoons and wetlands have also a recognised human dimension, and they constitute an invaluable historical and cultural heritage, e.g. the lagoon of Venice or the smaller coastal lagoons scattered along the Mediterranean coast.

Over the last twenty years, the scientific community has taken an increasing interest in these important areas. Since the symposium organised in 1981 by UNESCO/SCOR Consultative Committee on coastal systems (Lasserre & Postma, 1982), several studies were published, focusing on hydrology, biology and ecological classification criteria, as well as on the coastal management and conservation strategies of coastal wetlands (Mitsch & Gosselink, 2000). In the last decade, the main research fields were functional ecology and biogeochemistry, in respect of ecosystem alterations and buffering capacity (see as an example Kjerfve, 1994; Caumette et al., 1996; Schramm & Nienhuis, 1996; Viaroli et al., 2005).

It is now urgent to identify the influence of global change, from regional (e.g. eutrophication, degradation, erosion and loss of natural habitats) and local impacts (e.g. urbanisation, contamination, and tourism activities). Among others, an important aspect lies in the conservation of distinctive elements of wetland biodiversity in a global context, as opposed to conservation targets framed from too narrow or arbitrary national and regional boundaries.

Identification of proxies and climate-sensitive keystone species having a large impact on the rest of the community needs to be achieved in a timely manner in order to develop appropriate monitoring programmes. The initiative of establishing an international Global Terrestrial Observing System (GTOS) and particularly its coastal module would be largely beneficial in improving the capacity for detecting and predicting the effect of global climate change on coastal systems (GTOS, 2005).

The international Conference "CoastWetChange—Lagoons and Coastal Wetlands in the Global Change Context: Impacts and Management Issues" was organized in Venice, 26–28 April 2004, at the initiative of UNESCO and CORILA, to provide an interdisciplinary forum to share knowledge and experience of recent developments in wetland science and global change. The aim was to identify gaps, problems and successes in the integration of global change issues into lagoon and coastal wetland management. Based upon current scientific evidence, climate change will create novel challenges for coastal and marine ecosystems that are already stressed from human development, land-use change, environmental pollution, habitat alteration and loss. Venice was an emblematic venue for this meeting. Included in the World Heritage List established by UNESCO under the World Heritage Convention, Venice and its lagoon are a unique place internationally recognized as a "laboratory" for sharing and improving innovative technologies, developing knowledge in science and culture and for providing opportunities for intellectual exchange.

Most of the papers contributed to the conference have been published earlier (Lasserre et al., 2005). This volume comprises of 13 selected papers, of which five are reviews and eight primary research papers. A group of review papers, Section 1, analyses the main ecological and hydrogeomorphic features of coastal wetlands, with respect to climate change, including changes in sea level. Two papers address the importance of lagoons and wetlands as sentinel ecosystems for coastal observations of global change, and on the significance of flooding and ecological risk assessment applied to specific and more general situations. Finally, the research papers (Section 2) highlight our present understanding of the recent evolution of lagoons and coastal wetlands, including population dynamics, community succession, biogeochemical processes and pollution, key biological elements and related indicators.

Not every author has chosen the state-of-the-art approach. Some have preferred to concentrate on, from their point of view, crucial problems, which need further elucidation. A few give a detailed analysis and synthesis of the human-induced changes and rehabilitation measures. These differences are probably significant for our knowledge today, which is patchy in both space and depth.

Acknowledgements The "CoastWetChange" International Conference was sponsored and funded by UNESCO-ROSTE and CORILA, under the patronage of the Municipality and the Province of Venice. We are most grateful to UNESCO-ROSTE, Venice, in particular to its Director, Howard Moore, and to Philippe Pypaert for hosting the Conference at Palazzo Zorzi and for supporting the participation of scientists, in particular from developing countries and from Eastern Central Europe. The personnel of CORILA, in particular Frederic Brochier and Barbara Giuponi, were instrumental in assisting with all aspects of preparing, organising and publishing the Conference. This publication of the Conference Proceedings has received support from the Intergovernmental Oceanographic Commission (IOC) and UNESCO's Division of Ecological and Earth Sciences. Finally, we wish to thank the International Scientific Committee of the Conference and reviewers who were extremely helpful in the choice of invited participants and in the manuscript revision.

References

Caumette, P., J. Castel & R. A. Herbert, (eds), 1996. Coastal lagoon eutrophication and anaerobic processes (C.L.E.AN). Hydrobiologia 329: xxviii+225 p.

Crossland, C. J., H. H. Kremer, H. J. Lindeboom, J. I. Marshall Crossland & M. D. A. Le Tissier, (eds), 2005. Coastal fluxes in the Anthropocene. The IGBP series, Springer, Berlin, 231 p.

Eisenreich, S. J., 2005. Climate changes and the European water dimension. A report to the European Water Directors. EUR 21553 EN. IES-JRC, European Commission, Ispra, Italy, 253 p.

GTOS, 2005. Coastal GTOS. Strategic design and phase 1 implementation plan. GTOS report 36, Rome, FAO.

Kjerfve, B., 1994. Coastal Lagoon Processes. Elsevier Science Publishers, Amsterdam, xx + 577 p.

Lasserre, P. & H. Postma, (eds), 1982. Coastal Lagoons. Proceedings of the International Symposium on Coastal Lagoons, Bordeaux, France, 8–14 September 1981. UNESCO, IABO, SCOR. Oceanologica Acta, special issue, 461 pp.

Lasserre, P., P. Viaroli & P. Campostrini, (eds), 2005. Lagoons and coastal wetlands in the global change context: Impacts and management issues. Proceedings of the International Conference, Venice, 26–28 April 2004. UNESCO, ICAM Dossier 3.

Mitsch, W. J. & J. G. Gosselink, 2000. Wetlands. J. Wiley, New York.

Schramm, W. & P. H. Nienhuis, (eds), 1996. Marine benthic vegetation. Recent changes and the effect of eutrophication. Ecological Studies 123. Springer, Berlin.

Valiela, I., G. Collins, J. Kremer, K. Lajtha, M. Geist, B. Seely, J. Brawley & C. H. Sham, 1997. Nitrogen loading from coastal watersheds to receiving estuaries: new method and application. Ecological Applications 7: 358–380.

Viaroli, P., M. Mistri, M. Troussellier, S. Guerzoni & A. C. Cardoso, (eds), 2005. Structure, Functions and Ecosystem Alterations in Southern European Coastal Lagoons. Hydrobiologia 550: vii + 269 p.

Hydrobiologia (2007) 577:5–15
DOI 10.1007/s10750-006-0413-8

LAGOONS AND COASTAL WETLANDS

Broad-scale modelling of coastal wetlands: what is required?

Loraine McFadden · Tom Spencer · Robert J. Nicholls

Abstract A Wetland Change Model has been developed to identify the vulnerability of coastal wetlands at broad spatial (regional to global (mean spatial resolution of 85 km)) and temporal scales (modelling period of 100 years). The model provides a dynamic and integrated assessment of wetland loss, and a means of estimating the transitions between different vegetated wetland types and open water under a range of scenarios of sea-level rise and changes in accommodation space from human intervention. This paper is an overview of key issues raised in the process of quantifying broad-scale vulnerabilities of coastal wetlands to forcing from sea-level rise discussing controlling factors of tidal range, sediment availability and accommodation space, identification of response lags and defining the threshold for wetland loss and transition.

Guest editors: P. Viaroli, P. Lasserre and P. Campostrini.
Lagoons and Coastal Wetlands in the Global Change Context: Impacts and Management Issues

L. McFadden (✉)
Flood Hazard Research Centre, Middlesex University, Queensway, Enfield EN3 4SA, UK
e-mail: L.McFadden@mdx.ac.uk

T. Spencer
Cambridge Coastal Research Unit, Department of Geography, University of Cambridge, Downing Place, Cambridge CB2 3EN, UK
e-mail: tom.spencer@geog.cam.ac.uk

R. J. Nicholls
School of Civil Engineering and the Environment, University of Southampton, Highfield Southampton SO17 1BJ, UK
e-mail: r.j.nicholls@soton.ac.uk

Introduction

Coastal zones are currently experiencing intense and sustained environmental pressures from a range of natural, semi-natural and anthropogenic drivers (Mitsch & Gosselink, 2000). Increased resource use, environmental protection and the incorporation of social and equity issues into decision-making must evolve in the context of physical and ecological systems which show multi-scale dynamics and considerable uncertainties in likely response to near future environmental change (Poff et al., 2002; Morris et al., 2002). Both short-term and geological records show that coastal wetlands are particularly sensitive to change within the coastal zone (Allen, 2000; Schwimmer & Pizzuto, 2000; French & Spencer, 2002). Given such sensitivities, changes in wetland extent, position and type can be expected as accelerated sea-level rise increases forcing on wetland systems. Specific wetland loss mechanisms may include a range of natural processes, including edge erosion and retreat; internal dissection by the expansion of creek networks and surface ponds; changes in inundation frequency, waterlogging and in situ vegetative and root decay, and also human modification of marsh topography, sedimentology, ecology and hydrology (Mendelssohn & Morris, 2000). Within these contexts, this paper presents a new broad-scale wetland model which focusses upon the impact of relative sea-level rise on wetlands within the coastal zone.

Improving on earlier broad-scale assessments of wetland vulnerability (Hoozemans et al., 1993; Nicholls et al., 1999) and underpinned by a greatly improved global wetlands database (Vafeidis et al., 2004), the Wetland Change Model (i) provides a dynamic and integrated assessment of regional to global patterns of coastal wetland vulnerability and wetland loss; (ii) determines the ecological sensitivity of different wetland types to environmental forcing and the likelihood of transition to other wetland types and (iii) permits the assessment of the relative importance of sea-level rise, sediment supply and coastal protection measures in affecting wetland vulnerability. This model represents one module within the DIVA integrated assessment model for coastal areas (Dynamic Interactive Vulnerability Assessment)—developed within the EU-funded DINAS-COAST Project (Dynamic and Interactive Assessment of National, Regional and Global Vulnerability of Coastal Zones to Climate Change and Sea-Level Rise, www.dinas-coast.net). The DIVA tool has been designed to assess impact and vulnerability of the coastal zone to sea-level rise at regional to global scales and is driven by a set of internally consistent 'mid-term' (until 2100) scenarios of sea-level rise and socio-economic drivers of societal sensitivity to plausible impacts of accelerated sea-level rise and adaptive capacity (Hinkel & Klein, 2003). DIVA identifies coastal units that are particularly vulnerable to sea-level rise and adverse human interventions and allows for the evaluation of a range of response options (McFadden et al., in press).

Following the aim of the DINAS-COAST Project, the Wetland Change Model transforms a dynamic assessment of wetland vulnerability into patterns of wetland loss and transition. It seeks to capture the broad-scale response of wetlands to sea-level rise, integrating key drivers of wetland behaviour including human impacts such as dike construction or wetland nourishment (increasing sediment supply). This paper discusses key concepts raised in the process of modelling broad-scale wetland behaviour, underlining the problems of analysis at such spatial scales. Future developments are also considered, especially how this type of approach could be linked to other broad-scale monitoring efforts.

Broad-scale modelling of wetland behaviour

Modelling broad-scale wetland response to sea-level rise is important from a number of perspectives. In the first instance it strengthens our understanding of the mechanisms which control the behaviour of the wetland system as a large-scale unit within the physical landscape. Identifying 'hotspots' of wetland loss and a broad-scale assessment of levels of wetland vulnerability enables coastal managers and national organisations to make decisions on the best use of limited resources (Hammar-Klose & Thieler, 2001). Such modelling forms a basis from which effective plans can be developed to manage wetland change. In addition to this spatial dimension, broad-scale modelling is important to our understanding of long-term trajectories of future marsh behaviour. Important feedback mechanisms at longer-time scales (e.g. elevation/accretion relationships) mean that short-term measurements cannot be simply extrapolated to identify behavioural trends within a medium- to long-term temporal framework.

The Global Vulnerability Assessment (or GVA) and its subsequent revision provided the first worldwide estimate of both socio-economic and ecological implications of accelerated sea-level rise (Hoozemans et al., 1993; Nicholls et al., 1999). Based on a range of simple assumptions concerning rates of sea-level rise, subsidence and the response of the wetlands to sea-level forcing, the GVA gives a first-order perspective on wetland loss rates. However, the datasets have incomplete coverage, only three wetland types are considered, and wetland losses are only controlled by tidal range and accommodation space. While most calculations were conducted at a national scale, only results aggregated to a regional or global level could be considered valid (Nicholls et al., 1999).

Mass-balance models that focus on vertical adjustment of wetlands given accelerated sea-level rise have identified a number of controls on wetland response to environmental forcing factors: e.g. Severn Estuary, UK (Allen, 1990); North Norfolk coast, UK; Hut Marsh, Scolt Head Island (French, 1993); Venice Lagoon (Day et al., 1999); and wetlands of Louisiana, USA (Koch et al., 1990). Useful as these analyses are in defining the envelope of response, they only give a one-dimensional view of wetland-sea-level rise relations. Complex patterns of sedimentation mean that such models may not accurately represent the true sediment volumes required to enable such systems to keep pace with sea-level rise (French et al., 1995). Other studies have considered, and in some cases modelled, the landward retreat of saltmarshes under present, and expected near-future, rates of sea-level rise. Thus, for example, open coasts marshes in Essex, England (Harmsworth & Long, 1986; Reed, 1988), the marshes of the eastern Scheldt, Netherlands (Oenema & DeLaune, 1988) and salt marshes in the Gulf of Gabes, Tunisia (Oueslati, 1992) have provided a range of information on erosion and accretion along seaward marsh margins. In addition, it has been argued that floristically-rich upper marshes will disappear under the landward retreat of enclosing barriers (French, 1993). Most detailed studies of wetland loss of the type outlined above are typically local and relatively short-term in nature. Whilst such studies can be a useful means of calibration for broad-scale analysis, there is the significant problem of upscaling observations to the regional scale and longer time periods appropriate to modelling the broad-scale response of the system (Mitsch & Day, 2004). These problems have been addressed by the development of Landscape Simulation Models which are proving effective in assessing both the present and expected near-future distributions of wetland habitat types, taking into account both vertical and horizontal adjustments. Such models use hydrologic sub-models to distribute fluxes of water, nutrients and sediments over a grid of several thousand individual cells. Each cell incorporates a sub-model for plant production and soil formation which, alongside the hydrologic sub-model, determines the vegetation community. With changing environmental conditions, each cell is repeatedly interrogated by a 'habitat switcher' which resets the vegetation community if certain thresholds to inundation, soil chemistry and salinity are exceeded. Mapping expected environmental change in the Mississippi delta has been achieved in this way (Reyes et al., 2000; Martin et al., 2002). However, the computation effort required for this type of modelling approach precludes its current use as a widespread broad-scale tool for wetland analysis.

The Wetland Change Model presented within this paper seeks to engage with both levels of the current analysis of wetland performance identified above, identifying the key dynamics of wetland response emerging from small-scale analyses, and building a model which can then be tested, in part, against the modelling of changing wetland extent at the landscape scale. Fundamental to this characterisation is a conceptual model that defines the parameters that control wetland behaviour (Fig. 1). This paper examines the primary components of this conceptual model; in doing so, the challenges of broad-scale modelling are discussed.

Identifying environmental factors driving broad-scale wetland change

The DIVA Wetland Change Model, following earlier models, is based on the assumption that wetland response to external forcing such as sea-level rise involves both horizontal migration and vertical adjustment (Phillips, 1986; Nicholls et al., 1999; Allen, 2000). Vertical and horizontal changes may act independently of each other, but system behaviour must be considered as the synergistic response of both components. This integrated response of the system is modelled using three broad, yet critical, environmental forcing factors.

Fig. 1 Wetland Change Model

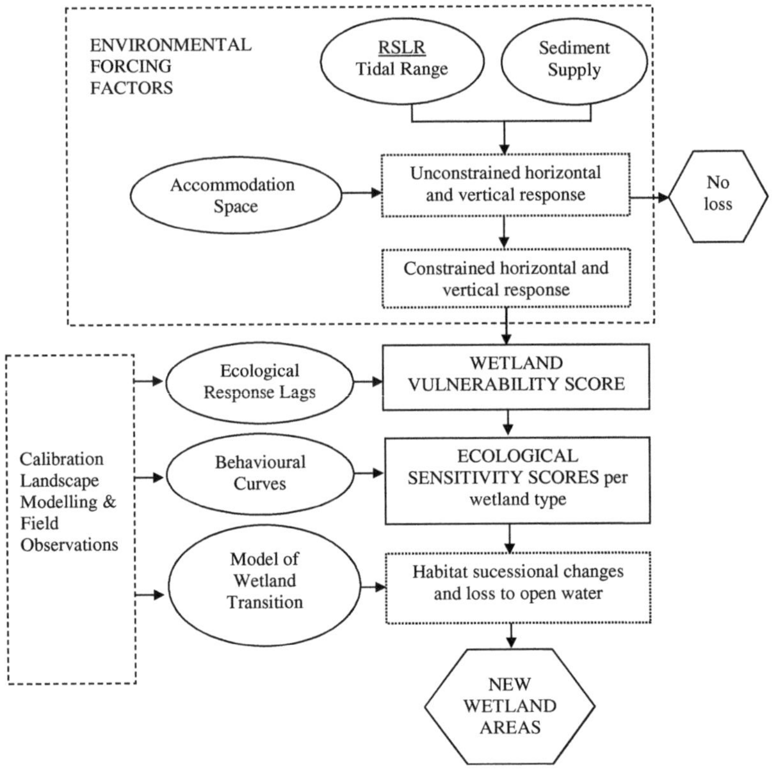

Ratio of relative sea-level rise to tidal range

A primary environmental forcing factor in driving vulnerability is the ratio of the rate of relative sea-level rise to tidal range. When sea-level rise is sudden and of high magnitude, as might result from sudden tectonic subsidence or high magnitude events such as tsunamis, a wetland may be completely submerged. Much more frequently, however, wetlands are subjected to slow rates of relative sea-level rise caused by eustatic factors and geological subsidence. Rather than submergence, the immediate impact of such gradual increases in sea level is a change in the nature of tidal flooding or hydroperiod (Reed, 1995). Hydroperiod is the cumulative inundation of surfaces due both to periodic flooding and to aperiodic tidal surge or high water levels associated with tidal surge or high river water flows and pulsed inputs of river sediments (Day et al., 1997). If wetlands are subject to a rise in relative sea-level without equal increases in elevation of the system, the duration and depth of tidal flooding will increase and communities can revert to a species composition typical of lower position in the tidal frame. In this situation tidal range is particularly significant in determining the vulnerability of the system to sea-level rise. It has been argued (Stevenson et al., 1986) that a wetland maintaining equilibrium under a large tidal range may have greater resilience towards the impacts of sea-level rise than a system existing within a narrower range of tidal fluctuation. As a result, modelling the combined impact of sea-level rise and tidal range is important in determining wetland response to sea-level forcing. Changes in storminess, direction of wave approach and tidal range are likely to accompany changes in mean sea level, but it is not possible to consider these effects in the current model framework.

Sediment supply

The long-term stability of coastal wetlands is also determined by the ability of wetland surfaces to maintain relative position in the tidal frame, thus keeping pace with the rate of sea-level rise (French, 1993). Regional trends in sediment supply are difficult to estimate due to their localised and highly variable temporal behaviour. There are often multiple sources of fine sediments (including riverine, cliff and offshore sources) on low-lying coasts and it is frequently difficult to isolate the contribution of particular sources, to assess the relative importance of local versus long-distance fine sediment transport and to differentiate between primary sediment supply and the re-mobilisation of previously transported sediments. In developed regions, human influences on the natural supply of sediment may significantly affect the response of wetlands over the long term. The submergence of Mississippi wetlands is partly due to the nature of catchment land management practices over the last 200 years that have reduced the supply of sediment to the inter-distributary bays. Similarly, more locally, coastal protection works often modify sediment transport pathways and sediment circulation systems.

A number of physical and human parameters are used within DIVA Wetland Change model to assess the impact of varying sediment supply on wetland vulnerability (Fig. 2). However, given the complexities of impact and response between sediment supply and wetland change, a comprehensive analysis of this forcing factor is not possible at the broad scale. A number of constraints on the model exist. Estimating the supply of a specific sediment type such as sand, mud, organic or inorganic, for example, cannot adequately be considered, so that only fine-grained sediment appropriate to the wetlands being studied can be assessed. Whilst it is clear that below-ground processes play an important role in coastal wetland stability (Nyman et al., 1995), the volume of sediment accreting on a wetland surface is the primary determinant of system response within the model. Sediment supply from in situ accumulation of organic sediments (Cahoon & Reed, 1995; Middleton & McKee, 2001; Rooth et al., 2003) or from external, inorganic inputs (French & Spencer, 1993; Christiansen et al., 2000) or a combination of the two, are used to characterise the impact of the environmental forcing factor within the DIVA model.

Accommodation space

The third driving factor is lateral accommodation space: given sufficient sediment supply to the system, this parameter is a key factor in determining the horizontal migration responses of wetlands. Coastal geomorphology has a major impact on accommodation space, where areas of high relief with steep coastal gradients reduce or remove the capacity for landward migration. Landward margins that have been fixed through coastal defence structures also effectively reduce the accommodation space, preventing horizontal migration.

Summary of environmental forcing factors

The Wetland Change Model combines environmental forcing on both horizontal and vertical response to give an assessment of the vulnerability of the total wetland area (Fig. 1). The model incorporates a number of physical (e.g. tidal range and sediment supply) and socio-economic forcing factors (e.g. removal of accommodation space by building seawalls and dikes). It is multidimensional in its characterisation of wetland vulnerability. It extends and refines the range of parameters that have been used in previous global assessments by taking account of all the main drivers of wetland change at broad scales. The model further builds on this characterisation by including a weighting component for each forcing

Fig. 2 Characterising sediment supply within the DIVA Wetland Change Model

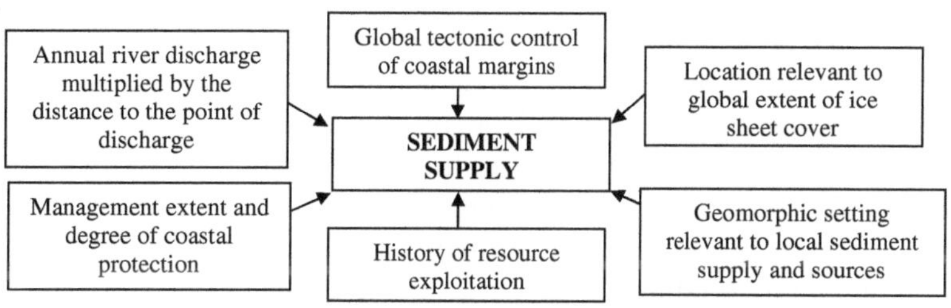

Table 1 The global weighting component for the environmental forcing factors

Ratio of relative sea-level rise to tidal range	0.5
Sediment supply	0.3
Accommodation space	0.2

factor (Table 1). The relative weighting of the environmental forcing factors reflect the importance of the parameter and the confidence with which it can be estimated at the broad-scale. This weighting component facilitates a greater resolution of system variability, recognising that each environmental forcing factor may exert a variable influence on wetland response depending on regional conditions.

Wetland response timescales

The response of a wetland to environmental stresses is not necessarily immediate. Rather, it is likely to be due to a combination of current and previous ecological states. This time lag between a forcing event and its geomorphological and/or ecological expression is dependent on habitat type. As a key aspect of the behaviour of wetlands to sea-level forcing, it is important that appropriate wetland response timescales are considered within broad-scale analyses. Incorporating such ecological lag time within the Wetland Change Model involves two conceptual developments: (1) global coastal wetland typology and (2) establishing relative response times for each wetland type.

Geographic variation in vegetation zonation has traditionally been used to form the basis for coastal wetland classifications, generally for establishing resource inventories and the identification of sites of particular conservation value. The refinement of this approach has been to use numerical techniques to establish differences in habitat type, e.g. on Argentinean marshes (Cantero et al., 1998) and on the Mississippi River deltaic plain (Visser et al., 1998). Such arguments have to some extent been driven by the Clementsian theory of deterministic, unidirectional change in ecosystem development (Clements, 1916) where plants are the primary drivers in trapping and binding sediments in intertidal environments and through determining elevation change, further control plant succession (and see Chapman, 1959 for a saltmarsh example). However, it is now clear that this is only one model for coastal classification, largely restricted to low-lying coasts with abundant sediment supply. Broader classifications for coastal mangroves for instance, have identified multiple categories for mangrove forests (Woodroffe, 1990) where geomorphical setting and the process environment differentiate between types. Such broad findings are also supported by research on the morphodynamics of tidally-dominated saltmarshes (Reed & French, 2001). The key to a robust classification of coastal types is therefore to establish the physical contexts within which different wetland types are found. This means that for the assessment of wetland vulnerability, a morphological classification (Woodroffe, 2002) into wetland settings and their structural/physical characteristics is of more value. Taking this view, six broad wetland types were identified as the basis of transition and loss within the Wetland Change Model (Table 2).

Building on this classification, various response times associated with each wetland type were determined. Table 3 outlines the continuum of response times which define ecological lag effects within the model. Many saltmarsh plant species, for example, can tolerate a wide range of inundation frequencies (and the variations in physical

Table 2 The classification of wetland type used within the Wetland Change Model

1. Coastal forested wetlands
2. Freshwater marsh
3. Saltmarsh
4. Mangrove
5. Unvegetated sediment > mean high water springs (sabkas)
6. Unvegetated sediment < mean high water springs (mud and sand flats)

Table 3 Relative response lags within the Wetland Change Model

Coastal forested wetlands	Highest response lag
Mangrove	↑
Freshmarsh/Saltmarsh	
Unvegetated sediment	Lowest response lag

and chemical soil characteristics which accompany them) and can rapidly colonise a range of new tidal habitats. By comparison, coastal forest tolerances are typically lower and colonisation of new habitat is difficult. For this habitat type, response will be strongly influenced by previous conditions, until a threshold point is reached when the system may collapse catastrophically (Cahoon et al., 2003). The relative response times of each wetland type were based on expert judgement combined with field observations. Incorporating response lag into the model transforms the assessment of the vulnerability of the total wetland area into a value of the ecological sensitivity of the six wetland types to sea-level rise (Fig. 1).

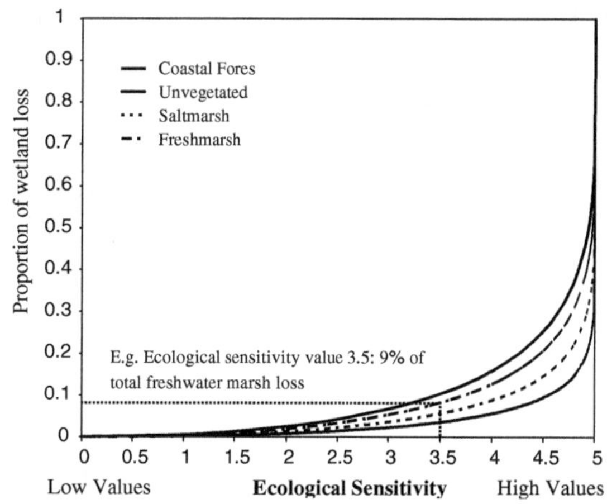

Fig. 3 Wetland loss, and wetland types as a proportion of total wetland loss, with changing wetland sensitivity (see text for explanation of ecological sensitivity)

Differentiation of wetland loss by wetland type

Existing large-scale models of wetland response to accelerated sea-level rise generally deal with the conversion of vegetated surfaces to open water and thus generate statistics on total loss of wetland area, e.g. GVA and subsequent revisions (Nicholls et al., 1999). Such models are most appropriate where local rates of relative sea-level rise are high, such as in subsiding, sediment-starved deltaic environments. However, under more moderate rates of sea-level rise and an adequate sediment supply ecosystem change may be (i) slower than predicted and (ii) involve change stepped across wetland types rather than simple loss, as ecological tolerances are exceeded in turn. The Wetland Change Model assesses both net wetland losses (due to conversion to open water) and transitions to other wetland types due to sea-level rise.

Linking the relative ecological sensitivities of wetland types to rates of wetland loss and transitions given sea-level rise requires (i) the construction of a series of wetland response curves (Fig. 3) which define the behaviour of the system by modelling the proportion of wetland expected to convert to another type given increasing exposure of a region to sea-level rise; and (ii) a model of wetland transition where loss is distributed between the wetland transitional types (Fig. 4).

Given the lack of information on broad-scale wetland behaviour, in the first instance both the wetland response curves and the transitional model were based on provisional estimates of wetland loss derived from expert judgement. Two primary datasets were used for calibration: (i) forecasting of changing wetland and open water areas in the Barataria and Terrebonne basins of South East Louisiana, USA from a basis of historical data collected by the United States Fisheries and Wildlife Service (USFWS) (D.J. Reed, pers. comm., 2003) and (ii) predictions of wetland type transitions produced by large-scale landscape modelling in the same region (Reyes et al., 2000). The Reyes model was initialised with the 1956 USFWS habitat map for the two basins and the results of a 32-year simulation compared against the 1988 map of the region (Reyes et al., 2000). Simulated maps showed a goodness-of-fit of 75% using a multiple resolution fit algorithm. The model was then run to the year 2018 under a range of scenarios.

The rate of increase in open water is a useful and readily definable summary measure of wetland loss. Table 4 shows the increase in the proportion of open water for the period 2000–2060 for four US Gulf Coast administrative units, calculated within the DIVA Wetland Change Model using the highest level of modelled sea-level forcing (1.07 m, 1990–2100) available from

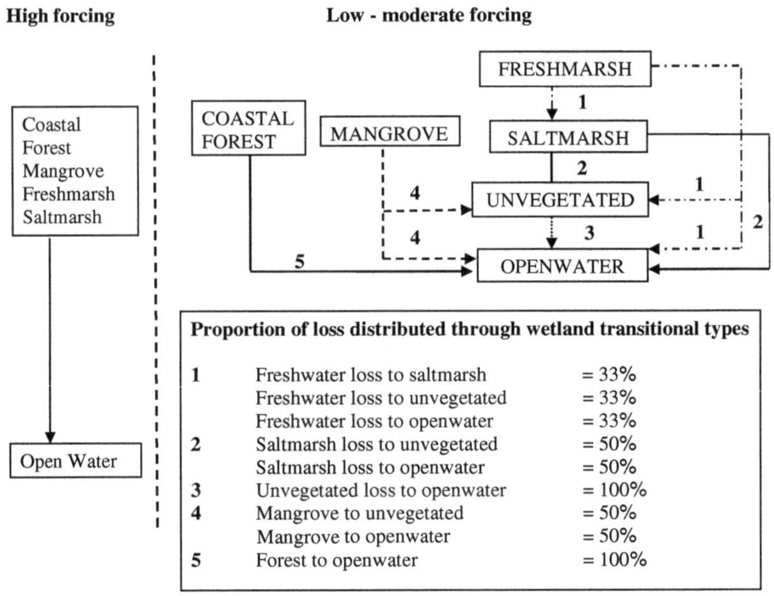

Fig. 4 Wetland loss and transitions between wetland types, to open water under sea-level rise: the Wetland Change Model

the Model. These data compare well with Reed's predictions of changes in the extent of open water in the Barataria and Terrebonne basins, with a similar timeframe and sea-level rise scenario.

The role of landscape modelling outputs (Reyes et al., 2000) within the DIVA calibration was two-fold. In the first instance, the results were used as a guide to the relative positions of the response curves within the envelope of vegetated wetland (Fig. 3). Outputs from the model were re-classified into the DINAS-COAST typology (Fig. 5) and basic trends in wetland loss were identified: the increase in open water at the expense of freshwater/brackish marsh and saltmarsh and the greater sensitivity of fresh marsh to sea-level forcing than saltmarsh within the basins. Less expected was the resilience of coastal forest which some authors (e.g. Conner & Day, 1988) have suggested might disappear from the Mississippi delta altogether, with continuous flooding preventing seedling establishment.

The landscape modelling data also provides some calibration of the point at which the model of wetland transition changes from gradual transition between types to complete submergence (Fig. 4). With lower forcing, transitions to other wetland types reflect gradual changes as salinity levels increase and environmental thresholds are crossed. At the present time, the model distributes wetland loss in equal proportions through the

Table 4 DIVA predictions of wetland conversion to open water in 4 US Gulf Coast States compared with predicted wetland/open water transition data for two basins in the Mississippi Delta (from Reed, pers. comm., 2003)

	DIVA Wetland Change Model parameters				Reed (pers. comm., 2003)	
	DIVA Administrative Units (Digital Chart of the World, ESRI, 2002)				Barataria	Terrebonne
	Texas	Louisiana	Alabama	Florida		
Coastal slope	Low forcing	Low forcing	Low forcing	Low forcing		
Tidal range	Low forcing	Low forcing	Low forcing	Low forcing		
Sediment supply	Moderate–high forcing	Moderate–high forcing	Moderate–high forcing	Moderate–high forcing		
Increase in open water 2000–2060	37%	26%	26%	32%	35%	23%

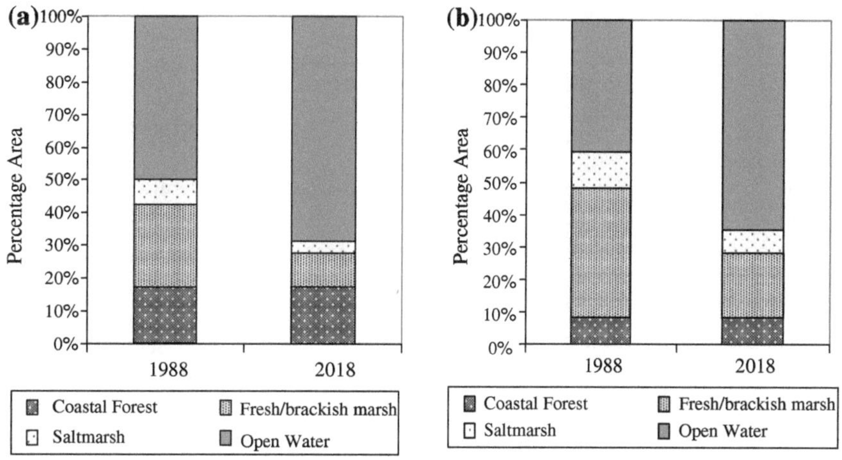

Fig. 5 Model outputs from Reyes et al. (2000) for the Barataria (a) and Terrebonne (b) basins, Mississippi delta, re-classified into the DINAS-COAST wetland typology

successive wetland types. However, under high levels of environmental forcing (high sea-level rise, low sediment supply and construction of barriers to horizontal wetland migration), the model converts all wetland losses to open water. The potential of the DIVA Wetland Change Model can be illustrated by the application of the model to another of the US Gulf Coast administrative units, the State of Florida (Fig. 6). The Model predicts an increase in open water from 2% in 2000 to 33% in 2060, largely at the expense of tidal flat environments but with some loss of saltmarsh and freshwater marsh. The resilience of coastal forest should be noted and that of mangrove forest, although as sea-level rise accelerates so mangrove areas begin to decrease.

The results from the DIVA Wetland Change Model appear commensurate with general estimates of global wetland losses given accelerated near-future sea-level rise. Nicholls et al. (1999), for example, have estimated that 22% of the world's wetlands could be lost by 2080 given a rise in global sea level of 38 cm. Table 5 shows the predicted loss of global wetlands over the time period 2000–2080 with low forcing scores for sediment supply and accommodation space under two sea-level rise scenarios. Although the model can predict regional to global vulnerability, a number of challenges remain, particularly when downscaling to regions where local effects may over-ride broad-scale controls. The development of more systematic national to regional scale assessments of wetland loss would further refine these estimates by contributing significantly to calibrating broad-scale models of the type presented here.

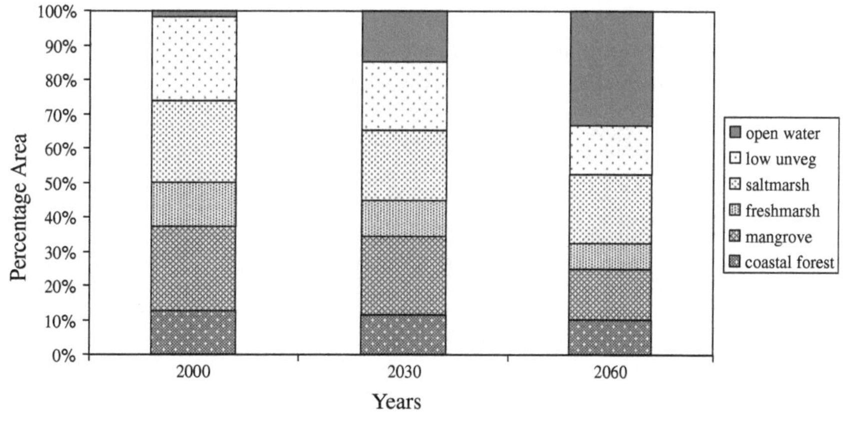

Fig. 6 Scenario of predicted wetland transitions 2000–2060 within the State of Florida, USA as predicted by the DIVA Wetland Change Model

Table 5 Predicted global wetland losses 2000–2080 under two sea-level rise scenarios

	Proportion of global wetlands lost, 2000–2080
RSLR—0.5 m (1990–2100)	
2020	0.10
2050	0.22
2080	0.32
RSLR—1 m (1990–2100)	
2020	0.17
2050	0.32
2080	0.44

Conclusions

The Wetland Change Model is an improved broad-scale model of loss and transition of coastal wetlands under sea-level rise. It incorporates all the primary drivers of wetland behaviour and provides an integrated perspective on the potential for wetland loss, examining a range of physical and human forcing parameters.

The major challenge lies in the validation of the model results. This is difficult at present due to the lack of suitable data and truly quantitative models of broad-scale wetland loss. The development of more systematic national to regional scale assessments of wetland behaviour would contribute significantly to validating the DIVA Wetland Change Model and hence, refining broad-scale estimates of wetland loss.

This work is in progress. As the model is applied and tested within DIVA it will inevitably be refined and improved. It is also hoped that this broad-scale modelling of coastal wetlands will stimulate improvements in, and extension of, field measurements of wetland behaviour, such that the data required to valid this type of model becomes more widely available.

Acknowledgements The authors gratefully acknowledge funding from the European Union under contract number EVK2-2000-22024. They thank all their partners in the DINAS-COAST project, particularly Z.B. Wang (Delft Hydraulics). We are grateful to Denise Reed, University of New Orleans, for directing us to the historical data on large-scale changing wetland and open water areas in the Barataria and Terrebonne basins, Mississippi delta, collected by the United States Fisheries and Wildlife Service (USFWS).

References

Allen, J. R. L., 1990. Salt-marsh growth and stratification: a numerical model with special reference to the Severn Estuary, southwest Britain. Marine Geology 95: 77–96.

Allen, J. R. L., 2000. Morphodynamics of Holocene salt marshes: a review sketch from the Atlantic and Southern North Sea coasts of Europe. Quaternary Science Reviews 19: 1155–1231.

Cahoon, D. R. & D. J. Reed, 1995. Relationships among marsh surface topography, hydroperiod and soil accretion in a deteriorating Louisiana salt marsh. Journal of Coastal Research 11: 357–369.

Cahoon, D. R., P. Hensel, J. Rybczyk, K. L. McKee, C. E. Proffitt & B. A. Perez, 2003. Mass tree mortality leads to mangrove peat collapse at Bay Islands, Honduras after Hurricane Mitch. Journal of Ecology 91: 1093–1105.

Cantero, J. J., R. Leon, J. M. Cisneros & A. Cantero, 1998. Habitat structure and vegetation relationships in central Argentina salt marsh landscapes. Plant Ecology 137: 79–100.

Chapman, V. J., 1959. Studies in saltmarsh ecology. IX. Changes in the saltmarsh vegetation at Scolt Head Island. Journal of Ecology 47: 619–639.

Christiansen, T., P. L. Wiberg & T. G. Milligan, 2000. Flow and sediment transport on a tidal salt marsh surface. Estuarine, Coastal and Shelf Science 50: 315–331.

Clements, F., 1916. Plant Succession: An Analysis of the Development of Vegetation. Carnegie Institute Publication 242, Washington DC.

Conner, W. H. & J. W. Day, 1988. Rising water levels in coastal Louisiana: implications for two coastal forested wetland areas in Louisiana. Journal of Coastal Research 4: 589–596.

Day, J. W., J. F. Martin, L. Cardoch & P. H. Templet, 1997. System functioning as a basis for sustainable management of deltaic ecosystems. Coastal Management 25: 115–153.

Day, J. W., J. Rybczyk, F. Scarton, A. Rismondo, D. Are & G. Cecconi, 1999. Soil accretionary dynamics, sea-level rise and the survival of wetlands in Venice Lagoon: A field and modeling approach. Estuarine Coastal and Shelf Science 49: 607–628.

Environmental Systems Research Institute (ESRI), 2002. Digital Chart of the World (DCW). ESRI, Redlands, California.

French, J. R., 1993. Numerical simulation of vertical marsh growth and adjustment to accelerated sea-level rise, North Norfolk, UK. Earth Surface Processes and Landforms 18: 63–81.

French, J. R. & T. Spencer, 1993. Dynamics of sedimentation in a tide-dominated backbarrier salt marsh, Norfolk, UK. Marine Geology 110: 315–331.

French, J. R. & T. Spencer, 2002. Sea level rise. In Warren, A. & J. R. French (eds), Habitat Conservation:

Managing the Physical Environment. J. Wiley, Chichester, 305–347.
French, J. R., T. Spencer, A. L. Murray & N. A. Arnold, 1995. Geostatistical analysis of sediment deposition in two small tidal wetlands, Norfolk, UK. Journal of Coastal Research 10: 308–321.
Hammar-Klose, E. S. & E. R. Thieler, 2001. Coastal vulnerability to sea-level rise: a preliminary database for the U.S. Atlantic, Pacific and Gulf of Mexico coasts. U.S. Geological Survey, Digital Data Series DDS-68.
Harmsworth, G. C. & S. P. Long, 1986. An assessment of saltmarsh erosion in Essex, England, with reference to the Dengie Peninsula. Biological Conservation 35: 377–387.
Hinkel, J. & R. J. T. Klein, 2003. DINAS-COAST: Developing a Method and a Tool for Dynamic and Interactive Vulnerability Assessment. IGBP LOICZ Newsletter, 27 (June 2003), 1–4.
Hoozemans, F. M. J., M. Marchand & H. A. Pennekamp, 1993. A Global Vulnerability Analysis: Vulnerability Assessment for Population, Coastal Wetlands and Rice Production on a Global Scale, 2nd edn. Delft Hydraulics, Delft, The Netherlands.
Koch, M. S., I. A. Mendelssohn & K. L. McKee, 1990. Mechanism for the hydrogen sulphide-induced growth in a Juncus-roemerianus marsh, west-central Florida. Journal of Coastal Research 11: 322–336.
Martin, J. F., E. Reyes, G. P. Kemp, H. Mashriqui & J. W. Day, 2002. Landscape modelling of the Missippi delta. Bioscience 54: 357–365.
McFadden, L., R. J. Nicholls, A. T. Vafeidis & R. S. J. Tol (in press). A methodology for modelling coastal space for global assessment. Journal of Coastal Research.
Mendelssohn, I. & J. T. Morris, 2000. Ecophysiological controls on the growth of Spartina alterniflora. In Weinstein M. A. & D. Kreeger (eds), Concepts and Controversies in Tidal Marsh Ecology. Kluwer, New York.
Middleton, B. A. & K. L. McKee, 2001. Degradation of mangrove tissues and implications for peat formation in Belizean island forests. Journal of Ecology 89: 818–828.
Mitsch, W. J. & J. W. Day, 2004. Thinking big with whole ecosystem studies and ecosystem restoration – the legacy of H.T. Odum. Ecological Modelling 178: 133–155.
Mitsch, W. J. & J. G. Gosselink, 2000. Wetlands. J. Wiley, New York.
Morris, J. T., P. V. Sundareshwar, C. T. Nietch, B. Kjerfve & D. R. Cahoon, 2002. Responses of coastal wetlands to rising sea level. Ecology 83: 2869–2877.
Nicholls, R. J., F. M. J. Hoozemans & M. Marchand, 1999. Increasing flood risk and wetland losses due to global sea-level rise: regional and global analyses. Global Environmental Change 9: S69–S87.
Nyman, J. A., R. D. DeLaune, S. Pezeshki & W. H. Jr. Patrick, 1995. Organic matter fluxes and marsh stability in a rapidly submerging estuarine marsh. Estuaries 18: 207–218.
Oenema, O. & R. D. DeLaune, 1988. Accretion rates in saltmarshes in the eastern Scheldt, southwest Netherlands. Estuarine, Coastal and Shelf Science 26: 379–394.
Oueslati, A., 1992. Salt marshes in the Gulf of Gabes (south eastern Tunisia): their morphology and recent dynamics. Journal of Coastal Research 8: 727–733.
Phillips, J. D., 1986. Coastal submergence and marsh fringe erosion. Journal of Coastal Research 2: 427–436.
Poff, L. F., N. M. Brinson & J. W. Jr. Day, 2002. Aquatic ecosystems and global climate change: potential impacts on inland freshwater and coastal wetland ecosystems. Pew Center for Global Climate Change, Arlington, VA.
Reed, D. J., 1988. Sediment dynamics and deposition in a retreating coastal salt marsh. Estuarine Coastal and Shelf Science 26: 67–79.
Reed, D. J., 1995. The response of coastal marshes to sea-level rise: survival or submergence? Earth Surface Processes and Landforms 20: 39–45.
Reed, D. J. & J. R. French, 2001. Physical contexts for saltmarsh conservation. In A. Warren & J. R. French (eds), Habitat Conservation: Managing the Physical Environment. J. Wiley, Chichester, 67–104.
Reyes, E., J. L. White, J. F. Martin, G. P. Kemp, J. W. Day & W. Aravamuthan, 2000. Landscape modeling of coastal habit change in the Mississippi delta. Ecology 81: 2331–2349.
Rooth, J. E., J. C. Stevenson & J. C. Cornwell, 2003. Increased sediment accretion rates following invasion by *Phragmites australis*: The role of litter. Estuaries 26: 475–483.
Schwimmer, R. A. & J. E. Pizzuto, 2000. A model for the evolution of marsh shorelines. Journal of Sedimentary Research 70: 1026–1035.
Stevenson, J. C., L. G. Ward & M. S. Kearney, 1986. Vertical accretion in marshes with varying rates of sea level rise. In Wolfe D. A. (ed.), Estuarine Variability. Academic Press, Orlando, 241–259.
Vafeidis, A. T., R. J. Nicholls, L. McFadden, J. Hinkel & P. S. Grashoff, 2004. Developing a Global Database for Coastal Vulnerability Analysis: Design Issues and Challenges. The International Archives of Photogrammetry, Remote Sensing and Spatial Information Sciences, Vol. XXXV Part B.
Visser, J. M., C. E. Sasser, R. H. Chabreck & R. G. Linscombe, 1998. Marsh vegetation types of the Mississippi River Deltaic Plain. Estuaries 21: 818–828.
Woodroffe, C. D., 1990. The impact of sea level rise on mangrove shorelines. Progress in Physical Geography 14: 483–520.
Woodroffe, C. D., 2002. Coasts: Form, Process and Evolution. Cambridge University Press, Cambridge.

Hydrobiologia (2007) 577:17–29
DOI 10.1007/s10750-006-0414-7

LAGOONS AND COASTAL WETLANDS

Coastal dynamics and wetlands stability. The Ebro delta case

Herminia I. Valdemoro · Agustín Sánchez-Arcilla · José A. Jiménez

Abstract Coastal wetland stability and structure can be significantly affected by littoral processes when they are close to the shoreline. Thus, under certain conditions, the combination of shoreline dynamics and direct wave action during storms can influence the stability of marshes and vegetation community composition. This interaction between littoral dynamics and coastal wetlands is illustrated by analysing processes taking place in the Buda Island (Ebro delta, NW Spain), where a coastal lagoon very close to a retreating shoreline exists. Two main time scales have been found to be relevant for interaction of coastal processes with ecosystem dynamics, the decadal and episodic scales. The decadal scale determines the average trend in beach width and directly controls the potential loss of wetland surface. The episodic scale is linked to the occurrence of wave and storm surge events and it determines a pulsing stress in the ecosystem through flooding, being key parameters to determine their direct influence the intensity and repetition of these events.

Keywords Ebro delta · Coastal erosion · Coastal wetlands

Introduction

Coastal wetlands are becoming a scarce commodity because of anthropogenic pressures and the failure our present society has had in recognizing the values and functions of these ecosystems. These values and functions include natural and socio-economic aspects and are only now beginning to be properly considered (e.g. Costanza et al., 1997). An example of the increasing interest for wetlands can be found in Carlsson et al. (2003), who performed an analysis on identification of attributes that increase and decrease the citizens perceived value of (constructed) wetlands, and where biodiversity was found to be one of two greatest contributors to welfare. In spite of this, human pressure on the coastal zone has steadily increased in such a way, that coastal wetlands have been decreasing in surface worldwide. To stop this process and to enhance strategies for ensuring or promoting the conservation of these ecosystems there is a need for studies to identify direct and indirect causes of

Guest editors: P. Viaroli, P. Lasserre & P. Campostrini
Lagoons and Coastal Wetlands in the Global Change Context: Impacts and Management Issues

Herminia I.Valdemoro (✉) · A. Sánchez-Arcilla ·
José A.Jiménez
Laboratori d'Enginyeria Marítima, ETSECCPB,
Universitat Politécnica de Catalunya,
c/Jordi Girona 1-3, Campus Nord ed. D1, 08034
Barcelona, Spain
e-mail: herminia.valdemoro@upc.edu

this surface loss (e.g. Edyvane, 1999; Boesch et al., 1994).

Some of the factors affecting wetland survival are linked to natural processes, being the unbalance between relative sea level rise (RSLR) and vertical accretion (e.g. Baumann et al., 1984; Chmura et al., 1992; Day et al., 1997) one of the most clearly identified potential threats. In this case, the "natural" factor can also feature a human-induced component since the unbalance is usually produced by a reduction of sediment supplies contributing to vertical wetland accretion due to the implemented water management policies. The importance of preserving "natural" river discharges to maintain the vertical elevation of wetlands (see e.g. Day et al., 1995; Hensel et al., 1998) has been clearly demonstrated. In any case, this approach implies to model wetland dynamics mainly as a vertical balance being the horizontal dimension seldom considered. With respect to the influence of the horizontal dynamics most of the studies have been done in tidal salt marshes with tidal currents being the main driving factors controlling the wetland evolution (see e.g. Allen, 1997, 2000; d'Alpaos et al., 2005 and references therein). Other works such as the Ruth and Pieper (1994) and Schwimmer & Pizzuto (2000) explicitly include the role of wave action in eroding exposed wetlands to modulate the response of wetlands to sea level rise (SLR).

Additionally, when coastal wetlands are close to the shoreline, it is clear that littoral processes can significantly affect wetland stability and, in that case, the horizontal dimension has to be added to the previously considered vertical analysis. A clear example of the effect of coastal processes in wetland stability is the role of barrier islands in protecting wetlands against wave and storm surge action. Considering the Louisiana coast, it has been in those areas where barriers islands have been eroded, that wetlands have been severely affected (Williams et al., 1997).

The influence of wave action on the stability of marshes and vegetation community composition and seagrass bed structure has been commonly analysed from the original works of Keddy (1982) through the so called Relative Exposure Index (REI) (see also Fonseca & Bell, 1998 and references therein). Other studies analysing the effects of wave action on vegetation composition in marshes, barriers and shorelines from different perspectives and environments are, among others, the ones due to Delaune et al. (1987), Roman & Nordstrom (1988), Fahrig et al. (1993), Costa et al. (1996), Courtemanche et al. (1999), Stallins & Parker (2003) and Roland & Douglass (2005).

To stress the potential influence of coastal dynamics on existing wetlands in Catalonia (NE Spain), Fig. 1 shows their spatial distribution. It can be seen that most of them, and especially those with a larger extension, are located in coastal areas. Among them, the largest number of coastal wetlands and the most important ones in terms of environmental values are located in the Ebro delta, a RAMSAR site, which is also part of the Natura 2000 Network.

Within this context, the paper will analyse the influence of littoral dynamics on coastal wetlands along the Ebro delta. To do this, a coastal lagoon very close to the shoreline in the Buda Island has been selected. The aim is to illustrate the interaction between this type of environments and the corresponding coastal fringe dynamics. From here the strong "physical" control on wetland dynamics will be presented which can be used to forecast wetland evolution and to assess impacts of human policies. The proposed approach must be thus considered as complementary to the one usually followed to analyse the wetland response to RSLR in terms of its vertical dynamics.

Study site

The Ebro delta, located 200 km southward of Barcelona in the Spanish Mediterranean coast, has an extension of 320 km^2 and a coastline 50 km long. It includes a Natural Park of 7,802 ha giving administrative protection to the areas of highest environmental value, including habitats like freshwater, brackish and saline lagoons, salt marshes and coastal and small sandy dune areas (Fig. 2).

The solid discharges of the Ebro river have been decreasing during the last decades and, as a consequence of this, the delta has become more influenced by wave action in such a way that it has been subject to an intense reshaping process

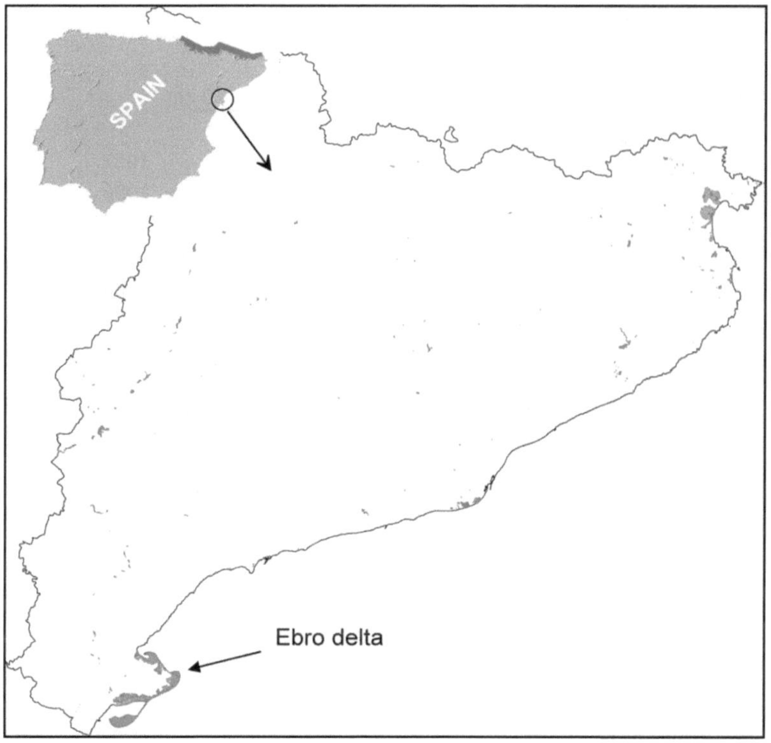

Fig. 1 Wetland distribution in Catalonia, NE Spain, (Department de Medi Ambient, 2004)

Fig. 2 The Ebro delta

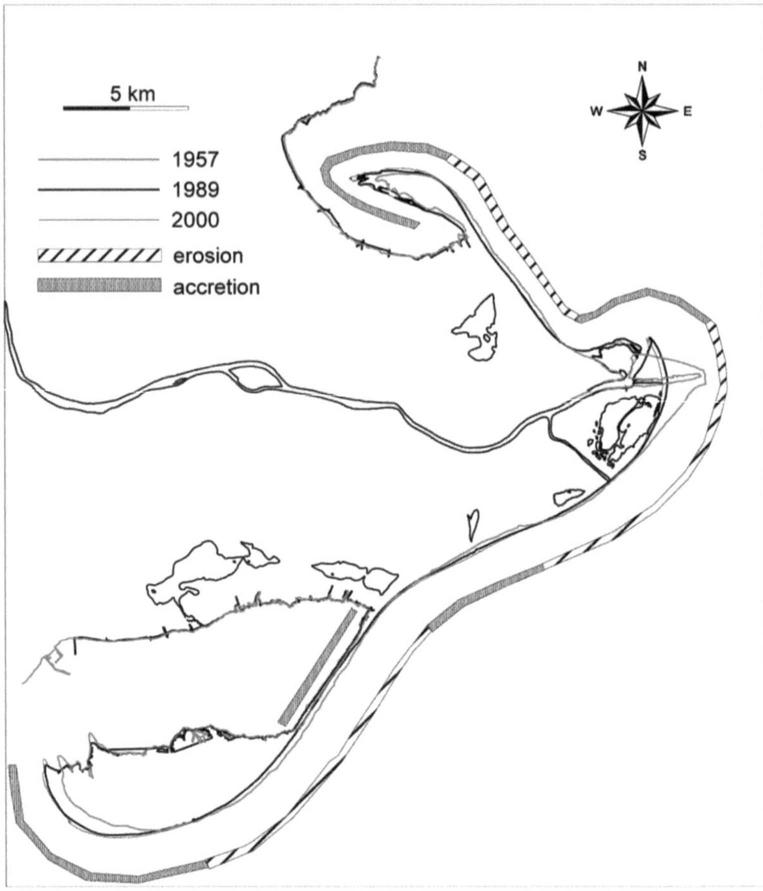

Fig. 3 Long-term shoreline evolution in the Ebro delta (modified from Jiménez & Sánchez-Arcilla, 1993)

(Jiménez & Sánchez-Arcilla, 1993; Jiménez et al., 1997). Figure 3 shows the erosion and accretion areas along the delta during the period from 1957 to 2000, where it can be seen that the largest erosion rates correspond to Cap Tortosa at the Buda Island area, whereas the apex of both spits has been experiencing a significant accretion. It can also be observed that the advance of the Northern spit towards the continent is tending to close the bay and, in that case, the process will finish by creating a new lagoon.

This cut-off in sediment supplies has led to a decrease in relative land elevation due to the scarcity of river floods able to distribute sediments into the deltaic plain. On the other hand, the last processed observations suggest, however, that the overall sub-aerial surface is nearly steady with a reshaping coastline as it was mentioned before. Further details on the Ebro delta coastal dynamics and shoreline evolution can be found in Jiménez & Sánchez-Arcilla (1993), Jiménez et al. (1997) and Guillén & Palanques (1997) among others.

A natural area of particular interest, Buda Island, is located in the Easternmost part of the

Fig. 4 The Illa de Buda area in the Ebro delta

delta and it has an approximate surface of 1,030 ha (measured from the year 2000 configuration) between the Migjorn river branch and the former Eastern river mouth (Fig. 4). About 350 ha of its surface located in the western part are dedicated to rice production. Most of the surface of Buda Island is occupied by Els Calaixos, a wetland of about 800 ha, from which about 350 ha correspond to a central lagoon with two basins. The remaining part corresponds to salt marshes, reed swamps, rushes and patches of Salicornia communities.

In Els Calaixos, the lagoon receives freshwater inputs from adjacent rice fields mainly during the period from May to October, whereas saltwater inputs mainly occur during winter when coastal storms are more frequent. This water input regime has generated two distinct domains: an inner basin in the W extreme which is mainly influenced by freshwater inputs and an outer one at the E border mainly influenced by storm induced water inputs (Comín et al., 1987). This salinity gradient has made this lagoon to be the unique wetland in the Ebro delta where macrophytes with different requirements such as *Ruppia cirrhosa*, *Potamogeton pectinatus y Zoostera noltii* coexist (Menéndez et al., 2002).

These authors have developed a functional model for primary producers in the lagoon as a function of hydrological changes. At the ends of the range of possible changes they establish that an increase in the freshwater inputs together with a decrease in the nutrient inputs should produce an increase in the spatial coverage of *P. pectinatus*. On the other hand, the increase of the influence of saltwater inflows should induce an increase in the coverage of *R. cirrhosa* together *Z. noltii*, specially in the E end of the lagoon. Due to this, and regardless of the ideal state of the lagoon in terms of ecosystem quality, it is clear that the dominance of one species over the other ones is strongly controlled by the balance between the two water inputs.

In addition to this, one of the most important vegetal communities of the area is the *Arthrocnemetum macrostachyum* and *Sarcocornia fruticosa* that presents one of the best preserved populations of Catalonia, which is included in the catalogue of habitats of natural interest at the European level. These communities together with those living in the sandy fringe between the lagoon and the sea are exposed to a significant threat due to the loss of physical substrate associated to the existing coastal erosion (see e.g. Departament de Medi Ambient, 1995). Further information about vegetation communities in the salt marshes of the area can be found in Curcó et al. (2002).

Wetland stability and coastal dynamics

The quantification of the wetland ecosystems change due to coastal processes can be achieved following two different approaches. The first one consists in analysing the ecosystem dynamics and estimating how they will be affected by coastal processes and/or driving factors such as incident waves in a detailed manner. This means to simulate all acting processes and quantifying the interactions between the physic and ecological components including the corresponding feedbacks. The second approach consists in applying some kind of parameterisation, which allows analysing the ecosystem in terms of a reduced number of variables and processes. With this approach, the final goal should be to characterise the major features of the targeted process and not the full interactions. In this work, we have followed the second approach and, to do that, we approach the problem at two different scales that are presented in what follows.

In both cases it is assumed that conservation, migration or erosion of the wetland can be parametrically modelled as a function of an external forcing such as sea level rise and wave action (see e.g. Schwimmer & Pizzuto, 2000; Mc Fadden et al., 2005). In most of the cases, the wetland stability is determined by its ability to keep in pace with the rate of sea level rise and to migrate landwards to compensate erosion in its outer end. In this sense, the presence of inland obstacles such as levees or dikes will reduce the accommodation space and will determine the wetlands to be more vulnerable to such external forcing. This is the case of the wetlands found in the Els Calaixos area (Fig. 4).

Wetland reduction due to shoreline erosion

The first considered interaction consists in the direct exposure of wetlands to coastal dynamics in areas where erosion processes dominate and, as a consequence of this, shoreline retreat will affect the wetland (see e.g. Valdemoro et al., 1999, 2001). The time scale associated to this interaction is determined by the dominant process driving coastal evolution.

In the Ebro case, systematic shoreline retreat along the Buda Island is mainly driven by the existence of alongshore gradients in longshore sediment transport rates, i.e. transport rates steadily increase along the coast (Jiménez & Sánchez-Arcilla, 1993). The time scale associated to this process is usually considered as yearly, i.e. the coastal evolution is controlled by the yearly-integrated net longshore sediment transport. However, the time scale to affect coastal wetlands stability is more related to the geomorphic response than to the agent itself. Thus, although the identified agent-scale is "yearly", its integrated action during a determined period should drive the interaction with the wetland. The minimum time scale for integration depends on the magnitude of the longshore sediment transport gradient on the one hand and on the beach width on the other hand. It is usually associated to decades, which is the main scale for the observed coastal reshaping processes (see e.g. Jiménez & Sánchez-Arcilla, 1993; Jiménez et al., 1997).

To illustrate this, Fig. 5 shows the evolution of wetland surface in Buda Island from 1957 to 2000. The shoreline evolution during this period is shown in Fig. 3 where a significant decrease of wetland surface is apparent, with a total loss of about 150 ha during the entire period.

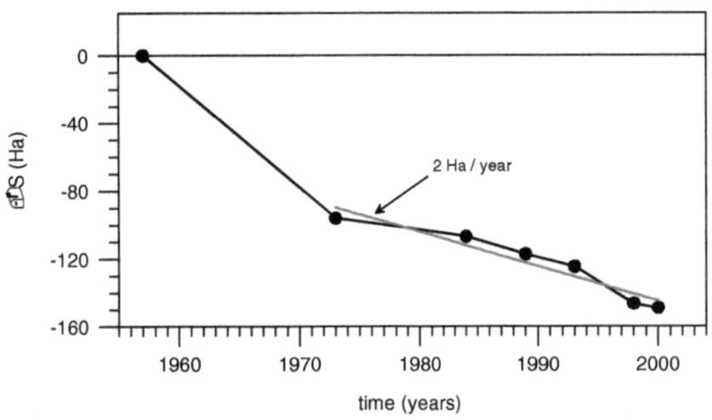

Fig. 5 Wetland surface evolution rate in Els Calaixos from 1957 to 2000

This surface loss has occurred at different rates, with the highest loss rates verifying during the period from 1957 to 1973, which is the span presenting the largest coastal erosion rates in the study area during the last decades (Jiménez & Sánchez-Arcilla, 1993). After this rapid loss, wetland surface slowly decreases at an averaged rate of about 2 ha/year with most of the losses verifying at the northern part of the area (see Fig. 6). In principle, this decrease in loss rates could be associated to the presence of lower erosion rates during this period. However, this is not strictly true, because along the entire Buda Island there are zones with lower retreat rates and other stretches experience erosion rates up to five times higher (Fig. 6).

Figure 7 presents a simple conceptual model to explain how the interaction between wetland and shoreline dynamic takes place. This model adapts the one presented by Jiménez & Sánchez-Arcilla (2004) for barrier beaches, and it also makes use of the critical width concept (Leatherman, 1979). This implies assuming that, in a retreating coast, wetlands located at a distance from the shoreline wider than a critical value do not experience surface loss because overwash does not reach the lagoon. This critical value depends on the actual wave and water level climates. However, when the shoreline erosion makes this distance narrower or equal to the critical value, shoreline erosion will be accompanied by a migration of the beach towards the lagoon and the outer end of the wetland will be buried by overwash deposits.

The resulting equilibrium distance between shoreline and wetland under present conditions (wave and water level climates) can be easily identified in Fig. 6, where it is observed that the northern extreme of the wetland is located at an averaged distance from the shore of about 160 m. This distance is consistent with that observed in averaged terms for the Trabucador barrier beach (Jiménez & Sánchez-Arcilla, 2004), as expected for coastal stretches subject to the same wave and water level climates.

This type of behaviour has some consequences to forecast the ecosystem evolution due to coastal

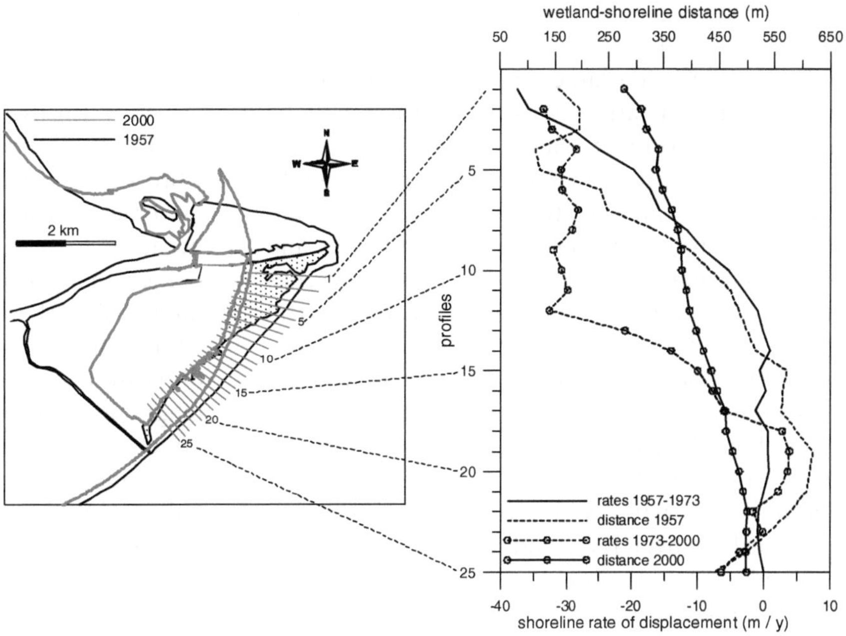

Fig. 6 Shoreline rate of displacement and distance of wetlands to the shoreline along the Buda Island from 1957 to 2000

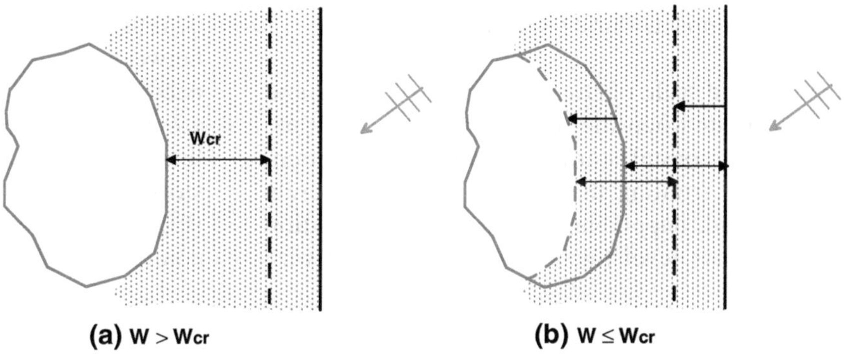

Fig. 7 Conceptual model of interaction between wetlands and shoreline dynamics (after Valdemoro, 2005)

dynamics. Menéndez et al. (2002) identified coastline retreat as the main factor for increasing the seawater dominance on local ecosystems. They associated the shoreline retreat with a reduction of the barrier beach separating the lagoon from the open sea. As a result, they forecasted an increase of the spatial distribution of sea grasses resistant to wave effects and a higher conductivity (e.g. *Zoostera noltii* or *Cymodocea nodosa*).

As it has been shown, the actual shoreline erosion does not necessarily imply a reduction of barrier excepting those areas where the actual width is wider than the critical value. However, in these last areas the reduction of the barrier will not affect wetlands unless it decreases below the critical width. On the other hand, in areas where this critical width has been already attained, the shoreline erosion does not imply a larger seawater influence since the barrier width in front of the wetland will remain stable in averaged terms.

Thus, the main direct influence of this shoreline erosion will be the loss of wetland surface as the barrier migrates landward. However, the input of seawater into the lagoon and its influence will remain constant provided the length of the coastal stretch with a width equal to the critical one does not vary in extension and for steady wave and water climates.

In the case that this length increases—as it is expected from the observed actual shoreline erosion rates which are increasing southwards

(Fig. 6), the potential influence of seawater on the wetland should increase because waves will be able to overwash a larger shoreline front and, in consequence, a larger amount of water will be able to get into the wetland.

In any case, if shoreline erosion rates significantly vary in time due to any natural or human-induced factors, the overwash transport rates and associated barrier response (in those stretches with a width equal or narrower than the critical value) will be affected. Using the model developed by Jiménez & Sánchez-Arcilla (2004) for barrier beach evolution, an increase in erosion rates along the coast will result in larger overwash transport rates and, in consequence, a narrower equilibrium barrier width together with larger landward rollover rates and a larger decrease in wetland surface. However, a decrease in shoreline erosion rates will induce lower overwash and rollover rates.

Wetland impacts during storms

The second scale of stress is the episodic one and it will occur when the wetland is affected by the impact of coastal storms. Under this condition, the beach fronting the wetland will be severely eroded and, depending on the relative elevation of the water level (including wave run-up) with respect to the barrier height, there will also be overwash. A conceptual model to predict the type of expected response in sandy barriers during extreme storms has been presented by Sallenger (2000). According to it, the usual situation of the Ebro delta coastal stretches under the impact of extremes storms will be overwash or inundation because, in general, no dunes high enough to prevent barrier/beach overtopping do exist. This implies that the impact of extreme storms in the Ebro delta will generally induce a landward flux of seawater that, depending on the overwash magnitude and the beach width can result in effects on the wetland.

Although the temporal scale is defined as episodic, it should be possible to assign a measurable time span for this interaction. This will be given by the return period of the smallest storm able to produce a "massive" flooding of the wetland and/or the full erosion of the emerged beach in front of the wetland. The return period is, thus, defined as the expected average interval between the occurrences of events above the necessary threshold. There are at least two main scenarios fulfilling the required conditions. The first one is associated to the presence of a wave storm without any significant meteorological tide (Fig. 8a). Under this situation the time scale is controlled by the return period of the wave

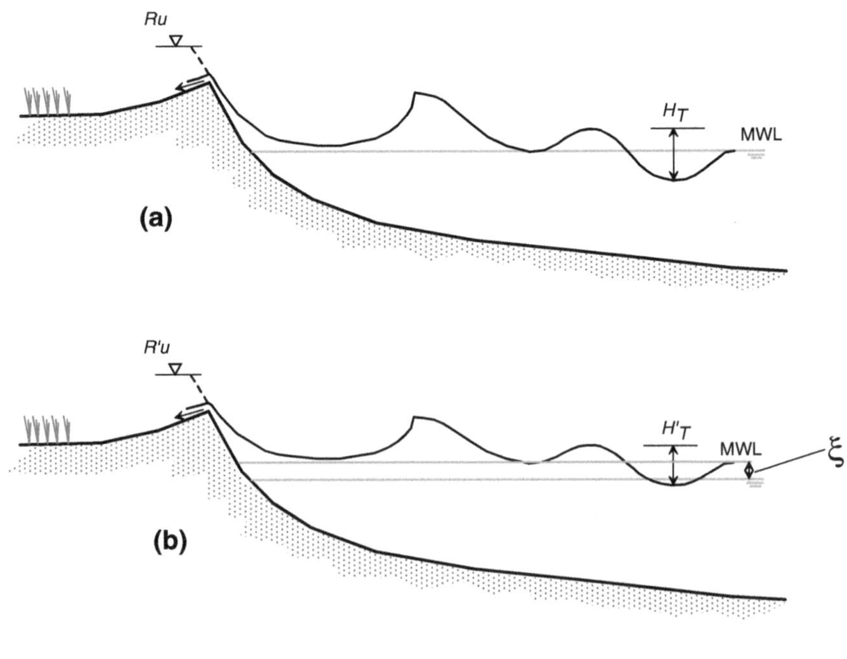

Fig. 8 Schematization of conditions affecting coastal wetlands during storm events. (**a**) Wave storm only. (**b**) Simultaneous wave and storm surge

height, H_T, during the storm to produce such response and it should be estimated from the extreme wave climate. The second one will occur when storms are characterised by the coexistence of high waves and a meteorological tide (Fig. 8b) and, the time scale will be related to the return period of the wave height-water level combination (joint probability). In this case, the minimum wave height H'_T should be linked to the storm surge, ξ in such a way that the higher the surge is, the smaller the wave height required to stress the system will be. In the specific case of the Ebro delta, the most energetic storm conditions are characterised by the presence of Eastern waves (llevants in vernacular), usually combined with the presence of a meteorological tide (see e.g. Jiménez et al., 1997).

Figure 9 shows two examples of the study area subject to the action of extreme events in which the interaction between the inner wetland and the sea is clearly seen. The first one corresponds to the beach morphology along the Buda Island after the action of a storm in October 1990 where some breaches can be observed. This storm was energetic enough to induce "massive" coastal erosion along the entire delta, including the breaching of the Trabucador barrier (see Sánchez-Arcilla & Jiménez, 1994 for further details). Hydrodynamic conditions during the storm can be summarised by a significant wave height of 4.5 m at the peak of the storm together with a meteorological tide of about 0.45 m (see details in Sánchez-Arcilla & Jiménez, 1994). These conditions together with a beach profile fronting the wetland without any protecting dune (the berm height measured before the storm was only 0.90 m above the mean sea level) determined the beach to be significantly overwashed. The estimated wave run-up at the peak of the storm will thus vary between a value of about 1.1 m—using the expression proposed by Stockdon et al. (in review)—and 1.8 m—using the expression of Mase (1989). In any case, if the storm surge at the peak of the storm is added, a total water level rise between 1.55 and 2.25 m should be obtained, which is clearly exceeding the berm height and fulfilling the conditions to produce massive overwash of the beach and, probably also the conditions to produce breaching (see e.g. Kraus & Wamsley, 2003).

These conditions created a connection between the lagoon and the open sea and, as a consequence, a continuous influx of seawater verified during the event. Moreover, after the pass of the storm the beach remained breached during several weeks and the connection between the lagoon and the sea was maintained. This resulted in a seawater inflow to the lagoon larger than expected because the breached conditions exceeded the duration of the storm event. Finally, the breach was artificially closed and the barrier morphology returned to the "normal" stage, i.e. the barrier equilibrium width.

The second situation corresponds to the action of a storm in October 1997 where a massive overtopping of the beach is clearly observed (Fig. 9b). Hydrodynamic conditions during the storm were a significant wave height of 4.9 m at the peak of the storm in absence of storm surge. The estimated wave run-up for these conditions varies between a value of about 1.4–2.30 m using the expressions of Stockdon et al. (in review) and Mase (1989), respectively. These values were significantly higher than those for the storm of October 1990 and they determined the full inundation of the beach.

It is clear from both examples that an important interaction between coastal dynamics and wetland properties should occur. It is expected that, due to its coastal location, this wetland ecosystem will be able to resist to this type of impulsive stress, the importance of such extreme events has to be carefully analysed. For instance, Gabriel & Kreutzwiser (2000) stated that system stability and recovery is dependent on a "normal" range of intensities, and frequencies. Moreover, they pointed out that stability can only be achieved if the total response time (sum of the relaxation and reaction times) is shorter than the interval between stress events. This clearly links the ability of the wetland to resist or to adapt to these events to the joint consideration of frequency and intensity values.

With respect to this last point, it has to be considered that overwash sites may be more susceptible to repeated high wave stress during high water levels, since they are generally at a

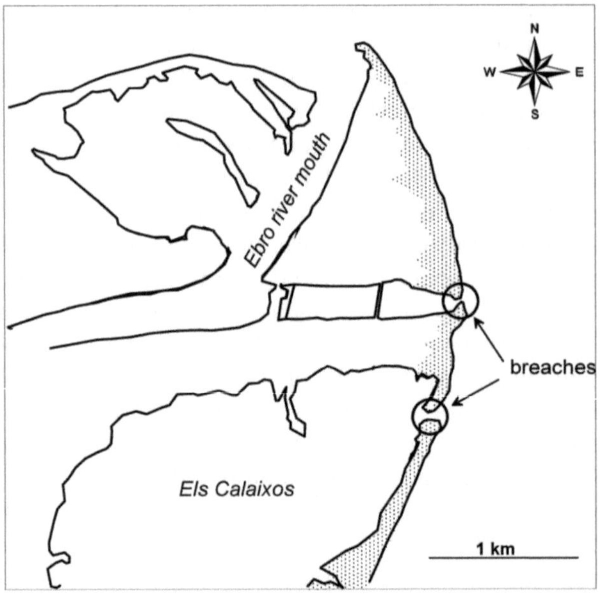

Fig. 9 Influence (impact) of extreme storms in the Els Calaixos wetland (Ebro delta). (**a**) Breachings after the impact of a storm in October 1990. (**b**) Inundation of the beach in Els Calaixos during a storm in October 1997

lower elevation. Moreover, when a large part of the beach fronting the wetland has been significantly eroded and it has attained the equilibrium beach width (Fig. 6), all this area will be easily overwashed during storms. The southern part of the area is also subject to significant erosion, which will make it evolve to a situation similar to the northernmost part, with overwash arriving to the wetlands. This is a synergic action between processes acting at episodic scale and those analysed at decadal scale.

Finally, with respect to the frequency of these events, Fig. 10 shows a time series of wave heights recorded off the Ebro delta from 1990 until 2004. The triangles in the figure mark events that have induced problems along the Ebro delta coast related to breaching and overwash. Here we refer to problems that have required some management actions to restore the post-storm situation back to the previous morphology. As it can be seen, at least during this climatic cycle, there is an increase in the frequency of problems associated to storm impacts. This could be due to the fact that during this period, the occurrence of extreme storms was higher (the winter of 2001–2002 and 2003–2004 can be considered as the

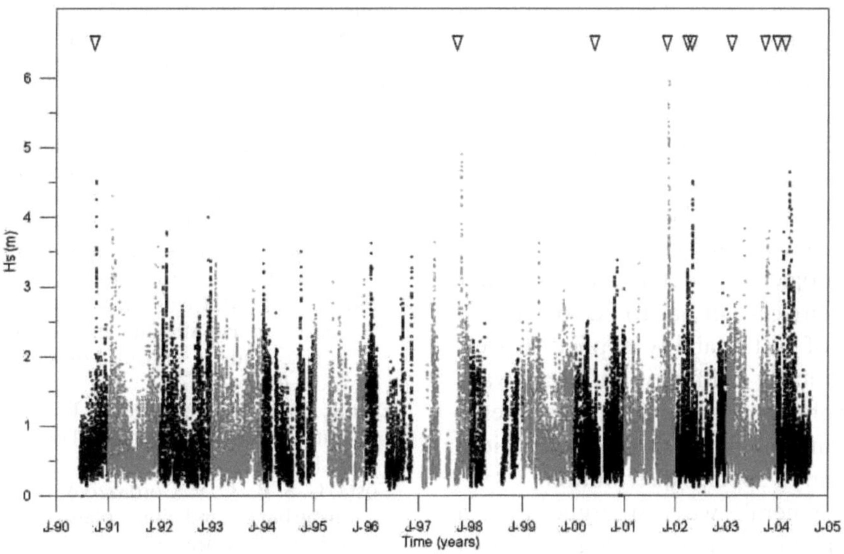

Fig. 10 Significant wave height time series recorded at 50 m depth off the Ebro delta from July 1990 to December 2004 (after Valdemoro, 2005). Triangles indicate events inducing problems related to breaching and/or inundation in some parts of the Ebro dleta coast

most stormy ones from the recorded time series in terms of intensity of storms and also in terms of extension of the storm season). It could also be due to the fact that storms are impacting in already affected (and, in consequence, more sensitive) areas.

Whether the combinations of these conditions and the ones presented in the previous section will affect the stability of the Els Calaixos wetland will largely depend on the frequency of occurrence and intensity of these events. Under natural conditions, it is not expected that the beach morphology will evolve so as to reduce their impacts. Moreover, since the freshwater input into the lagoon is artificially controlled, this can be used to modulate the influence of seawater in any direction (introducing more freshwater or reducing it) and, the type of ecosystem that will form the wetland.

Concluding remarks

Coastal wetlands are subject to a richness of processes that control water quality and the associated ecosystems. The morphodynamics of the fronting beach affects this water quality by controlling the exchanges with the sea through and above the fronting coastal fringe. This paper has shown how the wetland stability in terms of the morphological component can be parameterised in relatively simple "physical" terms. If this horizontal dynamics is combined with predictions of salinity and vegetation changes through the wetland, a more accurate description of the overall ecosystem changes should be obtained. In essence, this consists in the integration of processes mainly acting in the horizontal axis together with those acting in the vertical one.

The two main time scales for defining the ecosystem dynamics have been shown to be the decadal and episodic scales. The decadal scale, in which yearly processes such as longshore and cross-shore sediment transport play a dominant role in beach dynamics, determines the average trend in beach width and rollover rates and directly controls the potential loss of wetland surface. The observation of these physical parameters together with a control of the ecosystem properties should allow a more accurate parameterisation.

The episodic scale is linked to the occurrence of wave and storm surge events. It has been shown that the key parameters are the intensity and repetition of these events. This is because the recovery time needed by the coastal fringe to come back to its original state. This recovery is, in any anthropologically modified wetlands, supported often by management actions.

The combined long-term and episodic observations should, thus, lead to more robust parameterisation and eventually to a well calibrated ecosystem dynamics model. This should, in turn, allow a more efficient management of coastal wetlands and to objectively define/measure their sustainability. In any case, it has to be considered that the concept of sustainability applied to wetlands can be practically defined in different ways for both geomorphic and ecological components such as maintaining a given surface, elevation, primary production or species distribution.

Although, in general, we can argue that a sustainable management of wetlands is possible, there exist situations such as highly dynamic coasts where this is not likely to occur. A clear example is the case analysed here since the actual coastline evolution pattern and rates will determine a progressive wetland disappearance at decadal scale. If management options are designed at the local scale—a given wetland—to maintain its actual physical and ecological properties, it is questionable to consider this a sustainable management. However, if the overall deltaic system is considered, it should be possible to design a management policy where the disappearance of some wetlands will be compensated by the appearance/creation of new ones.

Acknowledgements This work was partially done in the framework of the FLOOD*site* project, funded by the European Commission under Contract No. GOCE-CT-2004-505420. The third author would like to thank the government of Catalonia for its support through the University Research Promotion Award for Young Researchers.

References

Allen, J. R. L., 1997. Simulation models of salt-marsh morphodynamics: Some implications for high-intertidal sediment couplets related to sealevel change. Sedimentary Geology 113: 211–223.

Allen, J. R. L., 2000. Morphodynamics of Holocene salt marshes: A review sketch from Atlantic and Southern North sea coasts of Europe. Quaternary Science Reviews 19: 1155–1231.

Baumann, R. H., J. W. Day & C. A. Miller, 1984. Mississippi deltaic wetland survival: Sedimentation versus coastal submergence. Science 224: 1093–1095.

Boesch, D. F., M. N. Josselyn, A. J. Mehta, J. T. Morris, W. K. Nuttle, C. A. Simenstad & D. J. P. Swift, 1994. Scientific assessment of coastal wetland loss, restoration and management in Louisiana. Journal of Coastal Research, Special Issue 20, 103 pp.

Carlsson, F., P. Frykblom & C. Liljenstolpe, 2003. Valuing wetland attributes: An application of choice experiments. Ecological Economics 47: 95–103.

Chmura, G. L., R. Costanza & E. Kosters, 1992. Modelling coastal marsh stability in response to sea level rise: A case study in Coastal Louisiana, U.S.A. Ecological Modelling 64: 47–64.

Comín, F. A., M. Menéndez & E. Forés, 1987. Salinidad y nutrientes en las lagunas del delta del Ebro. Limnética 3: 1–8.

Costa, C. S. B., C. V. Cordazzo & U. Seeliger, 1996. Shore disturbance and dune plant distribution. Journal of Coastal Research 12: 133–140.

Costanza, R., R. d'Arge, R. de Groot, S. Farber, M. Grasso, B. Hannon, K. Limbert, S. Naeem, R. V. O'Neill, J. Paruelo, R. G. Raskin, P. Sutton & M. van den Belt, 1997. The value of the world's ecosystem services and natural capital. Nature 387: 253–260.

Courtemanche R. P., M. W. Hester & I. A. Mendelssohn, 1999. Recovery of a Louisiana barrier island marsh plant community following extensive hurricane-induced overwash. Journal of Coastal Research 15: 872–883.

Curcó, A., C. Ibañez, J. W. Day & N. Prat, 2002. Net primary production and decomposition of salt marshes of the Ebre delta (Catalonia, Spain). Estuaries 25: 309–324.

D'Alpaos, A., S. Lanzoni, M. Marani, S. Fagherazzi & A. Rinaldo, 2005. Tidal network ontogeny: Channel initiation and early development. Journal of Geophysical Research-Earth Surface 110, 10.1029/2004JF000182.

Day, J. W., D. Pont, P. F. Hensel & C. Ibañez, 1995. Impacts of sea-level rise on deltas in the Gulf of Mexico and the Mediterranean: The importance of pulsing events to sustainability. Estuaries 18: 636–647.

Day, J. W., J. F. Martin, L. Cardoch & P. H. Templet, 1997. System functioning as a basis for sustainable management of deltaic ecosystems. Coastal Management 25: 115–153.

Departament de Medi Ambient, 1995. Pla de gestió de l'illa de Buda (Parc natural del delta de l'Ebre). Generalitat de Catalunya, Barcelona.

Delaune, R. D., S. R. Pezeshki & W. H. Patrick, 1987. Response of coastal plants to increase in submergence and salinity. Journal of Coastal Research 3: 535–546.

Edyvane, K. S., 1999. Coastal and marine wetlands in Gulf St Vincent, South Australia: Understanding their loss and degradation. Wetlands Ecology and Management 7: 83–104.

Fahrig, L., B. Hayden & R. Dolan, 1993. Distribution of barrier island plants in relation to overwash

disturbance: A test of life history theory. Journal of Coastal Research 9: 403–412.
Fonseca, M. S. & S. S. Bell, 1998. Influence of physical setting on seagrass landscapes near Beaufort, North Carolina, USA. Marine Ecology Progress Series 171: 109–121.
Gabriel, A. O. & R. D. Kreutzwiser, 2000. Conceptualizing environmental stress: A stress–response model of coastal sandy barriers. Environmental Management 25: 53–69.
Guillén, J. & A. Palanques, 1997. A historical perspective of the morphological evolution in the lower Ebro river. Environmental Geology 30: 174–180.
Hensel, P., J. W. Day Jr., D. Pont & J. N. Day, 1998. Short term sedimentation dynamics in the Rhone River Delta, France: The importance of riverine pulsing. Estuaries 21: 52–65.
Jiménez, J. A. & A. Sánchez-Arcilla, 1993. Medium-term coastal response at the Ebro Delta, Spain. Marine Geology 114: 105–118.
Jiménez, J. A. & A. Sánchez-Arcilla, 2004. A long-term (decadal scale) evolution model for microtidal barrier systems. Coastal Engineering 51: 749–764.
Jiménez, J. A., A. Sánchez-Arcilla, H. I. Valdemoro, V. Gracia & F. Nieto, 1997. Processes reshaping the Ebro delta. Marine Geology 144: 59–79.
Keddy, P. A., 1982. Quantifying within-lake gradients of wave energy: Interrelationships of wave energy, substrate particle size, and shoreline plants in Axe Lake, Ontario. Aquatic Botany 14: 41–58.
Kraus, N. C. & T. V. Wamsley, 2003. Coastal barrier breaching, Part 1: Overview of breaching processes. ERDC/CHL CHETN IV-56, US Army Engineer Research and Development Center, Vicksburg.
Leatherman, S. P., 1979. Migration of assateague Island, Maryland, by inlet and overwash processes. Geology 7: 104–107.
Mase, H., 1989. Random wave runup height on gentle slope. Journal of Waterway, Port, Coastal and Ocean Engineering 115: 649–661.
McFadden, L., T. Spencer & R. J. Nicholls, 2005. Identifying Vulnerable Wetland Systems: Modelling the Impact of Sea-level Rise on Large-scale Wetland Response. Proc Solutions to Coastal Disasters 2005. ASCE, Charleston, 453–465.
Menéndez, M., O. Hernández & F. A. Comín, 2002. Spatial distribution and ecophysiological characteristics of macrophytes in a Mediterranean Coastal Lagoon. Estuarine, Coastal and Shelf Science 55: 403–413.
Rolan, R. M. & S. L. Douglass, 2005. Estimating wave tolerance of Spartina alterniflora in Coastal Alabama. Journal of Coastal Research 21: 453–463.
Roman, C. T. & K. F. Nordstrom, 1988. The effect of erosion rate on vegetation patterns of an east coast barrier island. Estuarine, Coastal and Shelf Science 26: 233–242.
Ruth, M. & F. Pieper, 1994. Modeling spatial dynamics of sea-level rise in a coastal area. System Dynamics Review 10: 375–389.
Sallenger, A, 2000. Storm impact scale for barrier islands. Journal of Coastal Research 16: 890–895.
Sánchez-Arcilla, A. & J. A. Jiménez, 1994. Breaching in a wave-dominated barrier spit: The Trabucador bar (NE Spanish Coast). Earth Surface Processes and Landforms 19: 483–498.
Stallins, J. A. & A. J. Parker, 2003. The influence of complex systems interactions on barrier island dune vegetation pattern and process. Annals of the Association of American Geographers 93: 3–29.
Schwimmer, R. A. & J. E. Pizzuto, 2000. A model for the evolution of marsh shorelines. Journal of Sedimentary Research 70: 1026–1035.
Valdemoro, H. I., 2005. La influencia de la morfodinámica en los usos y recursos costeros. Ph. D. thesis. Universitat Politècnica de Catalunya, Barcelona.
Valdemoro, H. I., J. A. Jiménez & A. Sánchez-Arcilla, 1999. Vulnerability: Evaluating the consequences of coastal changes. An application to deltaic coasts. Coastal Sediments'99, ASCE: 1450–1461.
Valdemoro, H. I., J. A. Jiménez & A. Sánchez-Arcilla, 2001. Vulnerability of Wetlands to Coastal Changes. A Methodological Approach with Application to the Ebro delta. Ecosystem and Sustainable Development III. Advances in Ecological Science Series n. 10, WIT Press: 595–604.
Williams, S. J., G. W. Stone & A. E. Burruuss, 1997. A perspective on the Louisiana wetland loss and coastal erosion problem. Journal of Coastal Research 13: 593–594.

Hydrobiologia (2007) 577:31–40
DOI 10.1007/s10750-006-0415-6

LAGOONS AND COASTAL WETLANDS

Coastal and estuarine ecological risk assessment: the need for a more formal approach to stressor identification

Michael C. Newman · Yuan Zhao · John F. Carriger

Abstract Application of ecological risk assessment to coastal and estuarine systems is accelerating although it initially lagged behind applications to land and freshwaters. Broader spatial and temporal scales, and multiple stressor integration are appropriately being considered more frequently in all risk assessment activities. This expansion and integration is essential for coastal risk assessment. Because coastal assessments must deal with co-occurrence of several candidate stressors manifesting within broad spatial and temporal scales, wider use of formal methods for assessing causal linkages is needed. Simple Bayesian inference techniques are discussed here to demonstrate their utility in quantifying the belief warranted by available information. The applicability of Bayesian techniques is illustrated with two examples, possible causes of fish kills on the Mid-Atlantic US coast and possible causes of hepatic lesions in fish of Puget Sound (Washington, US).

Guest editors: P. Viaroli, P. Lasserre & P. Campostrini
Lagoons and Coastal Wetlands in the Global Change Context: Impacts and Management Issues

M. C. Newman (✉) · Y. Zhao · J. F. Carriger
Virginia Institute of Marine Science, College of William and Mary, Rt. 1208 Greate Rd., Gloucester Point, VA 23062, USA
e-mail: newman@vims.edu

Keywords Risk assessment · Multiple stressors · Causality · Bayesian inference · Decision-making

Introduction

Initially, application of ecological risk assessment to U.S. marine habitats lagged behind applications to freshwater and land. The reason was not that coastal resource assessment was less important. More than three quarters of all commercial and recreational fish and shellfish species depend on estuaries (Lewis et al., 2001) yet these valuable coastal habitats remain in serious trouble (U.S. Commission on Ocean Policy, 2004). A lack of legal mandates was not the reason: ample U.S. federal legislation existed (see Rand & Carriger, 2001). The reasons for delay seem to arise from cultural biases (Newman & Evans, 2002) and political boundaries. There was an historical delay in implementation as North Americans slowly began to question the assumption that the oceans were too vast to be impacted by humans. As an important example, coastal eutrophication research lagged two decades behind that addressing freshwater eutrophication (Arhonditsis et al., 2003). Historical use patterns of coastal resources tended to be more laissez-faire than use patterns established for land ownership and obligations associated with terrestrial landscapes. Relative to political boundaries, marine pollution issues are more likely to require interstate or international action for which issues of sovereignty and coordination delay decisions (Deacon et al., 1998).

Coastal habitats now receive an appropriate level of attention, and ecological risk assessment concepts framed initially with terrestrial or freshwater systems in mind are rapidly being adapted to marine systems. Ecological risk assessment, as currently practiced, requires some shift in emphasis to be most effective for marine systems. Crucial changes include expansion of the ecosystem context to encompass central ecotonal and landscape themes, e.g., Brown et al. (2002) and Kiddon et al. (2003). Some assessments of coastal systems can be compromised by not considering the appropriate spatial and temporal scale (Yanagi & Ducrotoy, 2003). Conservation action associated with tributyltin use is a good example of spatial shortcomings in coastal management. Terlizzi et al. (2004) concluded that marine protected areas (MPAs) do not protect *Hexaplex trunculus* populations from tributyltin-induced imposex, stating, "The most important reason for the limited biological effectiveness of MPAs is that the scale of processes in marine systems is often much larger than scales the reserve can encompass." Hawkins et al. (2002) also argue that the temporal scale applied to assessments of tributyltin is too short. Lastly, the co-occurrence of significant multiple stressors is more likely in coastal systems than in terrestrial or freshwater systems.

Consideration of broader scales and possible multiple stressors necessitates more integrated and formal identification of stressors (e.g., Brown et al., 2002; Munns et al., 2002). Qualitative approaches to identifying the most likely stressors from a suite of candidates in conventional assessments are being developed based on qualitative methods such as those of Hill (1965) or Fox (1991). The sufficiency of these different approaches remains untested for coastal assessments requiring vigilant determination of the likelihood that a candidate stressor is, in fact, the stressor needing attention. Being qualitative and often dependent on unstructured expert opinion, these approaches can be susceptible to common errors in human cognition and problem solving (Newman & Evans, 2002). Bayesian techniques provide more formal mathematics for causal assessments, and consequently, for reducing the likelihood of making such errors. Applying Bayesian methods can enhance decision-making.

Two examples illustrate this last point about formally identifying the most plausible stressor(s) in complex coastal systems. Recent controversy about fish kills notionally due to the dinoflagellate, *Pfiesteria piscicida*, demonstrates the high costs of causal model development without adequate consideration of alternate causes. Cancer prevalence in Puget Sound fish demonstrates Bayesian methods for isolating the most plausible cause from many possible causes.

Fish kills caused by *P. piscicida*

From Bacon's 1620 *Novum Organum* to recent cognitive theory (Piattelli-Palmarini, 1994), one thing is clear: the unaided human mind has limited abilities to accurately determine causality and likelihood. The minds of coastal risk assessors are no exception as can be illustrated with recent investigations of fish kills along the Mid-Atlantic U.S coast. In 1992, Burkholder et al. (1992) proposed that *P. piscicida* caused large fish kills in coastal North Carolina. Uncertainty about the cause of these and subsequent fish kills generated much debate among regional scientists and resource managers with early expert opinion favoring the hypothesis that excess nutrient-induced blooms of the toxin-producing *P. piscicida* caused the kills. Suggestions were made that *P. piscicida* blooms also pose a health hazard to humans contacting infested waters. When subsequent large fish kills occurred in Mid-Atlantic coastal waters, substantial revenue was lost in the associated States as seafood sales dropped and tourists avoided the region. Regional decision-making was compromised because the informal expert opinion process became mired in accusations of ethical misconduct, risk exaggeration, and legislative stonewalling (Newman & Evans, 2002; Whitehead et al., 2003; Belousek, 2004). How important *Pfiesteria* or low dissolved oxygen conditions were relative to causing fish kills remains in active area of debate and research.

Table 1 Hill's nine aspects of noninfectious disease association applied to causality assessment for fish kills notionally related to *P. piscicida*

Strength of Association	The probability of fish kill when *P. piscicida* is present is 20.5%, as opposed to 4.9% when it is not present (calculated from Burkholder et al., 1995; Newman & Evans, 2002). Numerous laboratory bioassays have shown that, when *P. piscicida* is present at a level high enough, the fish percent mortality is as high as 100%, as opposed to 0% in the control.
Consistency of association	North Carolina data indicate that there were 9 (Burkholder et al., 1992: Table 1) and 10 (North Carolina Division of Water Quality, http://h2o.enr.state.nc.us/esb/Fishkill/fishkillmain.htm) fish kills, in 1991–1992 and 1997–2003, respectively, associated with *Pfiesteria*.
Specificity of association	North Carolina fish kill data from 1997 to 2003 indicate that among 371 fish kills, only 10 were suspected to be related to toxic effect of *Pfiesteria*. About 93 were associated with low DO conditions. Other causes such as accidental toxic spills or field runoff may also have had significant roles. The specificity of association is low.
Temporal sequence	Among the fish kills linked to *Pfiesteria*, it is difficult to tell the order of occurrences of fish kills and high densities of *Pfiesteria*.
Biological gradient	In the reported field fish kill data, there's no apparent gradient between the number of fish killed and the exposure density and duration to *P. piscicida*.
Plausible biological Mechanism	There is no consensus about the mechanism of *P. piscicida* killing fish yet. One possible mechanism is toxin release. The structural information of the toxin has been only partially defined (Moeller et al., 2001) and it has been demonstrated to be present in extracts from *P. piscicida* strain (Burkholder & Glasgow, 2002). The other possible mechanism is micropredatory feeding (Vogelbein et al., 2002).
Coherence with general Knowledge	The relationship between *P. piscicida* and fish kills is coherent with the generally accepted knowledge that extensive algal blooms can cause fish death, though the mechanisms vary.
Experimental evidence	Considerable experimental evidence exists in the literature, indicating that *P. piscicida* can cause fish death.
Analogy	Most of the toxic dinoflagellates produce polyketide toxins (Miller & Belas, 2003). The toxin-generating mechanism described for *P. piscicida* is similar to this although its structure is still unclear. Its biological activity is often lost within a short period of time, while the typical toxins from other dinoflagellates are stable or can be easily stabilized (Moeller et al., 2001). On the other hand, the experiments with *Pfiesteria shumwayae* cultures of Vogelbein et al. (2002) suggested that the organisms caused fish mortality by micropredatory feeding, not exotoxin production.

In the presence of significant uncertainty about causal relationships, qualitative rules-of-thumb, such as Hill's nine aspects of disease association (Hill, 1965) and Fox's rules of practical causal inference (Fox, 1991), can guide judgments about plausibility of candidate causes of adverse effects. Table 1 illustrates the application of Hill's rules to fish kills notionally caused by *P. piscicida.* These rules are intended to foster a qualitative sense of plausibility for a candidate cause: they are not designed to rigorously compare candidate causes. More quantitative abductive methods are afforded by Bayesian statistics and are presented here as a means of improving causal inferences in marine systems.

Stow (Stow 1999, Stow & Borsuk 2003) used straightforward Bayesian methods to identify an influential, but misleading, inference about coastal North Carolina fish kills. *Pfiesteria piscicida* was found at sites of fish kills 17 of 33 times during three consecutive years of sampling, leading Burkholder et al. (1995) to conclude that "*P. piscicida* was implicated as the causative agent of 52±7% of the major fish kills ... on an annual basis in North Carolina estuaries and coastal waters." The flaw in this conclusion can be illustrated with Bayes's Theorem,

$$p(\text{Fish Kill}|\mathit{Pfiesteria}) = \frac{p(\text{Fish Kill})p(\mathit{Pfiesteria}|\text{Fish Kill})}{p(\mathit{Pfiesteria})}$$

In words, the probability of a fish kill occurring given *P. piscicida* was present is equal to the product of the probability of a fish kill occurring times the probability of finding *P. piscicida* if a fish kill did occur divided by the probability of finding *P. piscicida*. The data collected during the three years estimated

$p(Pfiesteria|\text{Fish Kill})$ but the conclusion was incorrectly made about $p(\text{Fish Kill}|Pfiesteria)$. Estimates of p(Fish Kill) and p(*Pfiesteria*) are needed to calculate the level of belief warranted about *P. piscicida* causing a fish kill. When Newman & Evans (2002) did this, the odds dropped from the stated 1:2 to 1:5 of a fish kill occurring when *P. piscicida* was present. The causal evidence was not as strong as originally suggested.

This basic approach can be extended to analyze two or more competing causal explanations in coastal systems, e.g., Borsuk et al. (2004). Here, it is extended to calculate the relative likelihood of two competing causes for fish kills being low dissolved oxygen (Low DO) versus *P. piscicida* at the time that Burkholder et al. (1995) made the causal influence about *P. piscicida*. The equation for calculating the probability of a fish kill given the presence of *P. piscicida* is provided above. That for low dissolved oxygen causing a fish kill is the following:

$$p(\text{Fish Kill}|\text{Low DO}) = \frac{p(\text{Fish Kill})p(\text{Low DO}|\text{Fish Kill})}{p(\text{Low DO})}$$

The two competing explanations can be expressed as the quotient of the two probabilities:

$$\frac{p(\text{Fish Kill}|Pfiesteria)}{p(\text{Fish Kill}|\text{Low DO})} = \frac{\dfrac{p(\text{Fish Kill})p(Pfiesteria|\text{Fish Kill})}{p(Pfiesteria)}}{\dfrac{p(\text{Fish Kill})p(\text{Low DO}|\text{Fish Kill})}{p(\text{Low DO})}}$$

$$= \frac{p(Pfiesteria|\text{Fish Kill})p(\text{Low DO})}{p(\text{Low DO}|\text{Fish Kill})p(Pfiesteria)}$$

The four probabilities needed to calculate this quotient can be estimated with existing data. As estimated by Burkholder et al., 1995, $p(Pfiesteria|\text{Fish Kill})$ is 0.52 based on the major fish kills in North Carolina coastal waters from 1991 to 1993. Estimation of p(*Pfiesteria*) can be expressed in two ways because field surveys report either densities of *P. piscicida* or *Pfiesteria*-like organisms (PLO). P(*Pfiesteria*) was 0.345 for PLO or 0.205 for *P. piscicida* (Newman & Evans, 2002).

The entire coastal North Carolina dissolved oxygen (DO) data set for the same period as the Burkholder et al. (1995) study (1/1/1991 to 12/31/1993) was retrieved from the EPA STORET database. The designation of low oxygen conditions used here was somewhat arbitrary but consistent with regulatory definitions. Solely for purposes of illustration, all DO values lower than 4.0 mg/l were considered as indicative of low DO conditions. There were 674 cases of low DO out of 7100 DO measurements; therefore, p(Low DO) was estimated to be 0.095.

Fish kill data from 1991 to 1993 in North Carolina Coastal waters and estuaries were obtained from Dr. Mark Hale, Division of Water Quality, North Carolina Department of Environmental and Natural Resources (personal communication). The fish kill event was categorized as co-occurring with low DO condition if there was a specific comment that low DO (generally less than 4.0 mg/l) was observed during the period of fish kill. This occurred for nine out of 41 cases so 0.220 is the estimated probability of the presence of low DO when fish kills happened (p(Low DO|Fish kill)).

In this illustration, the likelihood ratio of fish kills due to *P. piscicida* versus fish kills due to low DO concentration can be calculated based on the above estimates for PLO and *P. piscicida* data, respectively:

$$\frac{p(Pfiesteria|\text{Fish Kill})p(\text{Low DO})}{p(\text{Low DO}|\text{Fish Kill})p(Pfiesteria)} = \frac{(0.52)(0.095)}{(0.220)(0.345)} = 0.651$$

$$\frac{p(Pfiesteria|\text{Fish Kill})p(\text{Low DO})}{p(\text{Low DO}|\text{Fish Kill})p(Pfiesteria)} = \frac{(0.52)(0.095)}{(0.220)(0.205)} = 1.095$$

The results show that, when PLO is used in the calculations, the likelihood that *P. piscicida* was the cause of a fish kill was lower than that of low DO. When *P. piscicida* only was used in the calculations, the likelihood of *P. piscicida* being the cause was approximately the same as that of low DO. Again, after the application of simple Bayesian methods, the level of belief warranted by the data changed from that originally inferred in 1995.

Two points can be made from this example. The decision based on available fish kill data to increase resources and funding to address the *P. piscicida* issue, but not dissolved oxygen issues, was not optimally informed. Also, these analyses suggest that putting more funding into producing better assessments of $p(Pfiesteria|\text{Fish Kill})$, $p(Low\ \text{DO}\ |\text{Fish Kill})$, $p(\text{Low DO})$, and $p(Pfiesteria)$, and gathering information relative to Hill's rules-of-thumb would result in much more informed decision-making.

Liver cancer in a sentinel fish species

Bayesian tools are also applicable for assessing the most plausible cause of an observed effect in situations where there are many candidate causes. The causal assessment of liver cancer in a sentinel fish species of Puget Sound (Washington, USA) will be used here to illustrate this point.

Hill's (1965) rules-of-thumb can be used to qualitatively judge plausibility of a particular stressor causing an effect. Newman (2001) applied Hill's nine aspects of disease association to hepatic cancer prevalence in English sole from Puget Sound, generating the following conclusions. (1) *Strength of association* between sediment PAH concentrations and cancer prevalence generally enhanced belief that PAH contamination caused cancerous lesions. (2) The *consistency of the association* was high between cancer prevalence and PAH concentration. (3) Logistic regression incorporating many candidate contaminants suggested that the *specificity of the association* between cancer and PAH concentration was moderate to high. (4) Unfortunately, the long latency period between exposure and cancer manifestation did not allow the fourth aspect (*consistent temporal sequence* of exposure then effect manifestation) to be directly assessed; however, experiments did demonstrate the appearance of precancerous lesions after juvenile exposure to PAH. (5) There was a *biological gradient* with cancer prevalence increasing with increasing PAH concentration. (6) A *plausible mechanism* existed, i.e., P450-mediated production of free radicals that form DNA adducts. (7) The proposed causal link of PAH to liver cancer was *coherent with existing, general knowledge* of carcinogenicity. (8) *Laboratory evidence* was produced to support this causal link. (9) Many other *analogous situations* existed in the literature. This application of Hill's rules suggested that sediment PAH contamination was a likely cause of the liver cancers in English sole. However, many candidate causes were present and quantitative discrimination among candidate causes was not done rigorously. Hill's nine rules-of-thumb can be adapted to such purposes but are not designed specifically to discriminate among candidate causes.

Formal Bayesian techniques allow one to be more explicit in large-scale assessments with several potential causes of adverse effect (e.g., Jones, 2001) but Bayesian techniques are infrequently applied to coastal assessment. The recent work of Borsuk (Borsuk, 2004; Borsuk et al., 2003; 2004) is a notable exception. More studies such as those of Borsuk et al. are warranted for complex coastal assessments in which several possible causes exist or the likelihood is high that effects result from multiple causes.

A Bayesian network can be constructed at the beginning of such a study. Bayesian networks show probabilistic connections (i.e., lines or arcs) between variables (nodes). Figure 1 is an example of a network that could be developed for the work of Myers et al. (1998). In Fig. 1, each connection represents a probabilistic dependency between a parent and a child node, and is represented by a function that mathematically defines its dependence on the parent variables.

Uncertainty about probabilistic dependencies (i.e., arcs) can be represented and quantified in Bayesian networks. Bayes's theorem allows both model parameters and observations to be probabilistically distributed and random. Using a

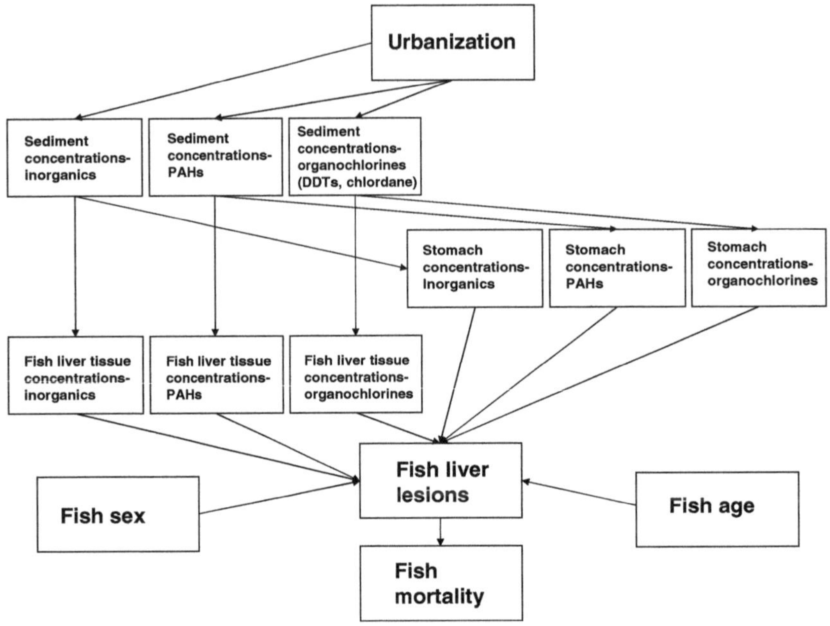

Fig 1 Bayesian belief network structured from the results of Myers et al. (1998). Arrows represent probabilistic dependencies and boxes represent variables

process called inversion, insight can be gained about causes using effects information (e.g., liver cancer prevalence and PAH): experimental observations (effects) may be used to infer the parameters (causes) of a probabilistic model (Robert, 1994). Conditioned on model parameters (causes), the future course a system takes may be described. The ability of Bayesian statistics to account for model uncertainty, by describing parameter uncertainty, makes them a powerful tool for coastal risk assessors. They have some drawbacks in developing uncertain model parameters for risk assessment as discussed by Aven & Kvaløy (2002).

The simple depiction of the nodes in Fig. 1 could be further detailed to reflect understanding of mechanisms and to reduce model uncertainty. For instance, nodes representing the various contaminant concentrations in different environmental compartments could be expanded into more specific chemical categories (e.g., organochlorines could have separate nodes for DDT compounds, dieldrin, chlordanes, PCB congeners) and hepatic lesions could be separated into lesion types (Myers et al., 1994).

Besides these expansions, additional data may cause shifts in the model functionality, making the network development an iterative process. For instance, Bayesian networks could be developed to reflect the statistical relationships for specific fish species because the appearance of lesions and relevance of different chemical classes in biotic or sediment compartments were found to vary among three studied fish species (Myers et al., 1994). These networks could take into account the relative risk for fish lesion prevalence from the output of the step-wise logistic regressions generated by Myers et al. (1990; 1994; 1998). Because Myers et al. (1994) conducted field studies to measure the strength of association of hepatic lesion prevalence in fish with different classes of contaminants, laboratory bioassays might be essential for updating and reinforcing the relationships in the network established from their field surveys. This was acknowledged in Horness et al. (1998), where a National Oceanic and Atmospheric (NOAA) database of sediment contaminant concentrations and hepatic lesion prevalence in English sole was used to set hypothetical sediment quality criteria. Although clear evidence correlates hepatic neoplasms and sediment contaminant concentrations, the association between tumor prevalence in fish and mortality was variable, and dose-dependence was not clear in some cases (Moore & Myers, 1994; Horness et al., 1998).

For management and predictive purposes, the interconnected nodes of a Bayesian model place multi-layered studies with many candidate causes (e.g., Myers et al. (1994)) under a single modeled framework. In creating a Bayesian network like the one in Fig. 1 for stressor assessment, it is important to include variables that may be changed through future interventions by management as was done in Borsuk et al. (2004). Of course, some variables such as fish age and sex (Fig. 1) are not amenable to manipulation by risk managers, but nonetheless, are pertinent as the state of such variables influence model outcomes. From the results of Myers et al. (1994), the age variable was an important determinant for the presence of some lesions. Sex did not have a significant relationship and a weaker probabilistic dependency would reflect this. Additional nodes with greater relevancy to remedial action by management could include loadings of specific contaminants from point and non-point sources. Resource managers can use such a model to set goals for criteria to protect fish species and can manipulate nodes to determine the implications from management activities or the sensitivity of various components in the model. In Fig. 1, the nodes representing PAH and organochlorine concentrations in sediment would have a stronger correlation with fish hepatic cancer. Variables regulating these concentrations would be the focus of future interventions. Easily implemented dependency analysis can help determine Bayesian network structure and foster implementation with available data sets (e.g., Cheng et al., 2002).

Bayesian networks are also conducive to linking data from separate studies into a cohesive model. In the studies of fish cancer and PAH from Myers et al. (1990; 1994; 1998) and Horness et al. (1998), where several years of data were gathered to identify potential causal agents, the incorporation of prior information is especially suited to this task. The frequentist methods used in Myers et al. (1990; 1994; 1998), and Horness et al. (1998) generally focus on the results from a single relevant study and attempt to objectively analyze those data. In the construction of prior probabilities, Bayesian methods can quantify results from previous experiments and combine them with those from a current experiment, even if they were conducted under dissimilar conditions (Spiegelhalter et al., 1999). For instance, previous studies cited in Myers et al. (1994) found lesions associated with exposure to contaminants including laboratory bioassays with field collected English sole and extracts of PAH from site-specific contaminants. Such information can be useful in establishing a prior distribution that does not rely on subjective information. Subjective probabilities can be used if there is a lack of such knowledge. In situations such as these, Borsuk et al. (2001) and Berry et al. (2003) implemented Bayesian hierarchical methods for cross-system meta-analyses to generate parameter estimates to specific systems that were information poor.

In addition to aiding causal assessment, Bayesian networks can contribute to risk management, remediation, and other aspects of risk assessment. Within a Bayesian network, nodes can reflect processes likely to influence an outcome, or how a management decision might change processes or outcomes. Three types of nodes can be used for these purposes: chance, decision, and utility nodes (Bacon et al., 2002). Chance nodes define processes related to the system; perhaps representing processes or states that affect or are affected by restoration activities. For instance, an assessment-derived conceptual model might be built within a Bayesian network framework that included probabilities associated with pathways of contaminant fate and transport amenable to restoration. Decision nodes are nodes representing potential decisions by the risk manager. Belief networks can contain nodes representing decision variables as well as decision constraints or criteria (Varis, 1997). The iterative procedure used in risk assessments allows for interim goals and changes in the Bayesian network structure can reflect the need for interim goals and decisions as they arise. Also, in the risk assessment process, assessment endpoints and measurement endpoints can serve as one of many foundations for criteria useful to decision nodes in a network.

Utility nodes are nodes that represent the satisfaction or gain that might accrue from decisions. To normalize cost/benefit variables across aesthetic and economic values, scales and weights can be assigned as was done in Bacon et al. (2002) for land management options and by Fenton and Neil (2001) to increase or decrease the importance of certain utilities. Different remediation techniques have various turn-around times, effectiveness, risks, and costs that can be included in utility nodes for a network. Along with listing the possible remediation alternatives, the strength of each alternative can be factored into the utility nodes of the network.

Concluding remarks

Optimal coastal and estuarine ecological risk assessment requires that the ecological risk assessment process be expanded to allow more integration of potential stressors and to include wider spatial and temporal scales. Fortunately, the means and impetus now exist for this to occur. The informal expert opinion approach, even when guided by sound rules-of-thumb, can be insufficient, as more candidate causes require consideration and wider scales are assessed. Simple Bayesian concepts and tools will be essential to effective risk assessment in the immediate future. The number of publications applying Bayesian statistics has increased in epidemiological and environmental journals, providing more examples to adapt in coastal risk assessments. Increased computation power and the availability of software such as Analytica, Netica (Norsys), or WinBUGS makes implementation of these methods easier. The explicit form and ability to express probabilities for plausible causes will accelerate decision making and remediation because, when probabilities are clear, judgments are more accurate and individuals are more willing to act (Keynes, 1921; Ellsberg, 1961; Chow & Sarin, 2001). Hopefully, application of methods such as those of Borsuk, Stow and Reckhow will become more routine and allow more effective coastal resource decision-making and action.

References

Arhonditsis, G., M. Eleftherisdou, M. Karydis & G. Tsirtsis, 2003. Eutrophication risk assessment in coastal embayments using simple statistical models. Marine Pollution Bulletin 46: 1174–1178.

Aven, T. & J. T. Kvaløy, 2002. Implementing the Bayesian paradigm in risk analysis. Reliability Engineering and System Safety 78: 195–201.

Bacon, P. J., J. D. Cain & D. C. Howard, 2002. Belief network models of land manager decisions and land use change. Journal of Environmental Management 65: 1–23.

Belousek, D. W., 2004. Scientific consensus and public policy: the case of *Pfiestera*. Journal Philosophy, Science & Law 4: 1–33.

Berry, D. A., S. M. Berry, J. McKellar & T.A. Pearson, 2003. Comparison of the dose-response relationships of 2 lipid-lowering agents: a Bayesian meta-analysis. American Heart Journal 145: 1036–1045.

Borsuk, M. E., 2004. Predictive assessment of fish health and fish kills in the Neuse River estuary using elicited expert judgment. Human and Ecological Assessment 10: 415–434.

Borsuk, M. E., D. Higdon, C. A. Stow & K. H. Reckhow, 2001. A Bayesian hierarchical model to predict benthic oxygen demand from organic matter loading in estuaries and coastal zones. Ecological Modeling 143: 165–181.

Borsuk, M. E., C. A. Stow, & K. H. Reckhow, 2003. Integrated approach to total maximum daily load development for Neuse River estuary using Bayesian probability network model (Neu-BERN). Journal of Water Resources Planning and Management July/August, 271–282.

Borsuk, M. E., C. A. Stow & K. H. Reckhow, 2004. A Bayesian network of eutrophication models for synthesis, prediction, and uncertainty analysis. Ecological Modeling 173: 219–239.

Brown, B. S., W. R. Munns Jr. & J. F. Paul, 2002. An approach to integrated ecological assessment of resource condition: the mid-Atlantic estuaries as a case study. Journal of Environmental Management 66: 411–427.

Burkholder, J. M. & H. B. Glasgow Jr., 2002. The life cycle and toxicity of *Pfiesteria piscicida* revised. Journal of Phycology 38: 1261–1267.

Burkholder, J. M., H. B. Glasgow Jr. & C. W. Hobbs, 1995. Fish kills linked to a toxic ambush-predator dinoflagellate: distribution and environmental conditions. Marine Ecology Progress Series 124: 43–61.

Burkholder, J. M., E. J. Noga, C. H. Hobbs, H. B. Glasgow Jr. & S. A. Smith, 1992. New "phantom" dinoflagellate is the causative agent of major estuarine fish kills. Nature 358: 407–410.

Cheng, J., R. Greiner, J. Kelly, D. Bell & W. Liu, 2002. Learning Bayesian networks from data: An information theory-based approach. Artificial Intelligence 137: 43–90.

Chow, C.C. & R.K. Sarin, 2001. Comparative ignorance and the Ellsberg Paradox. Journal of Risk and Uncertainty 22: 129–139.

Deacon, R. T., D. S. Brookshire, A. C. Fisher, A. V. Kneese, C. D. Kolstad, D. Scrogin, V. K. Smith, M. Ward & J. Wilen, 1998. Research trends and opportunities in environmental and natural resource economics. Environmental and resource Economics 11: 383–397.

Ellsberg, D., 1961. Risk, ambiguity, and the Savage Axioms. The Quarterly Journal of Economics 75: 643–669.

Fenton, N. & M. Neil, 2001. Making decisions: Using Bayesian nets and MCDA. Knowledge-Based Systems 14: 307–325.

Fox, G. A., 1991. Practical causal inference for ecoepidemiologists. Journal of Toxicology and Environmental Health 33: 359–373.

Hawkins, S. J., P. E. Gibbs, N. D. Pope, G. R. Burt, B. S. Chesman, S. Bray, S. V. Pround, S. K. Spence, A. J. Southward & W. J. Langston, 2002. Recovery of polluted ecosystems: the case for long-term studies. Marine Environmental Research 54: 215–222.

Hill, A. B., 1965. The environment and disease: association or causation? Proceedings of the Royal Society of Medicine 58: 295–300.

Horness, B. H., D. P. Lomax, L. L. Johnson, M. S. Myers, S. M. Pierce & T. K. Collier, 1998. Sediment quality thresholds: estimates from hockey stick regression of liver lesion prevalence in English sole (*Pleuronectes vetulus*). Environmental Toxicology and Chemistry 17: 872–882.

Jones, R. N., 2001. An environmental risk assessment/management framework for climate change impact assessments. Natural Hazards 23: 197–230.

Keynes, J. M., 1921. A Treatise on Probability. Macmillan, London.

Kiddon, J. A., J. F. Paul, H. W. Buffum, C. S. Strobel, S. S. Hale, D. Cobb & B. S. Brown, 2003. Ecological condition of U.S. mid-Atlantic estuaries, 1997–1998. Marine Pollution Bulletin 46: 1224–1244.

Kiker, G. A., T. S. Bridges, A. Varghese, T. P. Seager & I. Linkov, 2005. Application of multicriteria decision analysis in environmental decision making. Integrated Environmental Assessment and Management 1: 95–108.

Lewis, M. A., G. Scott & S. J. Klaine, 2001. Marine and estuarine toxicology and chemistry. Environmental Toxicology and Chemistry 20(1): 1–2.

Miller, T. R. & R. Belas, 2003. *Pfiesteria piscicida*, *P. shumwayae*, and other *Pfiesteria*-like dinoflagellates. Research in Microbiology 154: 85–90.

Moeller, P. D. R., S. L. Morton, B. A. Mitchell, S. K. Sivertsen, E. R. Fairey, T. M. Mikulski, H. B. Glasgow, N. J. Deamer-Melia, J. M. Burkholder & J. S. Ramsdell, 2001. Current progress in isolation and characterization of toxins isolated from *Pfiesteria piscicida*. Environmental Health Perspectives 109(supplement 5): 739–743.

Moore, M. J. & M. S. Myers, 1994. Pathobiology of chemical associated neoplasia in fish. In Malins D. C. & G. K. Ostrander (eds), Aquatic Toxicology: Molecular, Biochemical and Cellular Perspectives. CRC/Lewis Publishers, Boca Raton, FL: 327–386.

Munns, W. R. Jr., W. J. Berry & T. H. DeWitt, 2002. Toxicity testing, risk assessment, and options for dredged material management. Marine Pollution Bulletin 44: 294–302.

Myers, M. S., L. L. Johnson, T. Hom, T. K. Collier, J. E. Stein & U. Varanasi, 1998. Toxico-pathic hepatic lesions in subadult English sole (*Pleuronectes vetulus*) from Puget Sound, Washington, USA: Relationships with other biomarkers of contaminant exposure. Marine Environmental Research 45: 47–67.

Myers, M. S., J. T. Landahl, M. M. Krahn, L. L. Johnson & B. B. McCain, 1990. Overview of studies on liver carcinogenesis in English sole from Puget Sound: evidence for a xenobiotic chemical etiology: I. Pathology and epizootiology. Science of the Total Environment 94: 33–50.

Myers, M. S., C. Stehr, O. P. Olson, L. L. Johnson, B. B. McCain, S.-L. Chan & U. Varanasi, 1994. Relationships between toxicopathic hepatic lesions and exposure to chemical contaminants in English sole (*Pleuronectes vetulus*), starry flounder (*Platichthys stellatus*), and white croaker (*Genyonemus lineatus*) from selected marine sites on the Pacific Coast, USA. Environmental Health Perspectives 102: 200–215.

Newman, M. C., 2001. Population Ecotoxicology. John Wiley & Sons, Chicester, UK.

Newman, M. C. & D. A. Evans, 2002. Enhancing belief during causality assessments: cognitive idols or Bayes's theorem? In Newman M. C. & M. H. Roberts Jr., R. C. Hale (eds), Coastal and Estuarine Risk Assessment. CRC/Lewis Press, Boca Raton, FL, 73–96.

Piattelli-Palmarini, M., 1994. Inevitable Illusions. How Mistakes of Reason Rule Our Minds. John Wiley & Sons, New York.

Rand, G. M. & J. F. Carriger, 2001. U.S. environmental law statues in coastal zone protection. Environmental Toxicology and Chemistry 20: 115–121.

Robert, C. P., 1994. The Bayesian Choice: A Decision-Theoretic Motivation. Springer-Verlag, New York.

Spiegelhalter, D. J., J. P. Myles, D. R. Jones & K. R. Abrams, 1999. Methods in health service research: an introduction to bayesian methods in health technology assessment. BMJ 319: 508–512.

Stow, C. A., 1999. Assessing the relationship between *Pfiesteria* and estuarine fish kills. Ecosystems 2: 237–241.

Stow, C. A. & M. E. Borsuk, 2003. Enhancing causal assessment of estuarine fishkills using graphical models. Ecosystems 6: 11–19.

Suter, G. W. II, W. R. Munns & J. Sekizawa, 2003. Types of integration in risk assessment and management, and why they are needed. Human and Ecological Risk Assessment 9: 273–279.

Terlizzi, A., A. L. Delos, F. Garaventa, M. Faimali & S. Geraci, 2004. Limited effectiveness of marine protected areas: imposex in *Hexaplex trunculus* (Gastropoda, Muricidae) populations from Italian marine reserves. Marine Pollution Bulletin 48: 164–192.

U.S. Commission on Ocean Policy. 2004. Preliminary Report of the U.S. Commission on Ocean Policy. http://oceancommission.gov/documents/prelimreport/welcome.html#full.

Varis, O., 1997. Bayesian decision analysis for environmental and resource management. Environmental Modelling & Software 12: 177–185.

Vogelbein, W. K., V. J. Lovko, J. D. Shields, K. S. Reece, P. L. Mason, L. W. Haas & C. C. Walker, 2002. *Pfiesteria shumwayae* kills fish by micropredation not exotoxin secretion. Nature 418: 967–970.

Whitehead, J. C., T. C. Haab & G. R. Parsons, 2003. Economic effects of *Pfiesteria*. Ocean & Coastal Management 46: 845–858.

Yanagi, T. & J.-P. Ducrotoy, 2003. Toward coastal zone management that ensures coexistence between people and nature in the 21st century. Marine Pollution Bulletin 47: 1–4.

Hydrobiologia (2007) 577:41–53
DOI 10.1007/s10750-006-0416-5

LAGOONS AND COASTAL WETLANDS

Observations on future sea level changes in the Venice lagoon

Davide Zanchettin · Pietro Traverso · Mario Tomasino

Abstract 'Venice is sinking while the sea level is rising' is a common statement in issues concerning the future of the Venice lagoon. The search for a reliable interpretative tool for measured sea-level changes has taken on more urgency since the sea-level rise was indexed as the consequence of global warming—with catastrophic scenarios for both the ecotone and the city, linked to increasing lagoon erosion, sudden modifications of biological equilibriums, loss of wetlands, salt aggression and an increasing frequency of exceptional high tide events. However, the peculiar hydrodynamics of the northern Adriatic Sea, made more complex by the freshwater inflow from the Po River, and the conceptual limits of existing long-term predictive systems, would suggest a more cautious approach to the scenarios yet proposed for the next century.

Keywords Sea level · Lagoon · Tidal flats · Subsidence · Climate · Venice · Po River · Northern Adriatic

Guest editors: P. Viaroli, P. Lasserre & P. Campostrini
Lagoons and Coastal Wetlands in the Global Change Context: Impacts and Management Issues

D. Zanchettin (✉) · P. Traverso · M. Tomasino
Environm. Sc. Dept., University of Venice,
Dorsoduro 2137, 30123 Venice, Italy
e-mail: davidoff@unive.it

Introduction

The Venice lagoon (Fig. 1) is situated along a low-lying coast within the easternmost boundary of the Po Plain and connected to the northern Adriatic Sea through three wide mouths (Chioggia, Malamocco and Lido). It is the largest lagoon in the Mediterranean since it extends for about 550 km^2. Given the enormous amount of (ecological, historical and economical) interest in both the ecotone and the city, the lagoon has undergone a number of anthropic interventions since the fifteenth century, in an attempt to preserve a state of unstable equilibrium by counteracting its natural evolution. These include diverting river outflows to outside the lagoon and opening and widening the tidal inlets (an historical introduction to human interventions in the Venice lagoon can be found in: Ravera, 2000). In this context, the best known and most debated symptom of the disruption of this fragile ecotone's delicate equilibrium is the periodic '*acqua alta*' phenomenon (excessive high tides with water at flood level). The city is in such a unique setting—in the centre of the lagoon and built on piles—that even storm surges with comparatively small amplitudes (of less than 1 m) can cause flooding. However, the search for a reliable tool for interpreting measured sea levels has taken on more urgency in recent decades when an increase of the city's susceptibility to these high tides and flooding

Fig. 1 Map of the northern Adriatic Sea showing the Venice lagoon and the Po River Delta

events has been observed (Canestrelli et al., 2001) and the global sea-level rise, that is expected to lead to progressively higher background water levels in the lagoon, was indexed as the consequence of global warming (Church et al., 2001).

The potential consequences of an increase in the mean sea level go beyond the scenario of the serious degradation of the city of Venice, as a sea level rise would impact on the entire ecosystem that surrounds the city, mainly in terms of the loss of valuable ecotope areas such as the intertidal flats (see "Sea level rise in the Venice lagoon: ecological impacts").

The Venice lagoon ecosystem is morphologically definable as estuarine, hence it is controlled substantially by tidal excursions (with a micro- to meso-tidal regime). The inflow of saltwater throughout the whole basin occurs by means of an intricate network of canals, through which tidal currents can propagate. In some inner areas (i.e. shoals inside the lagoon) where waves are actually more important than the tides for the hydrodynamics (for instance in terms of sediment entrainment), the 'tidal breath' induced by the inflow/outflow through the lagoon mouths also assumes a fundamental ecological role by controlling the emersion/submersion ratios in marshy areas (*barene*) through changes in sea-level height.

As for the potential impact of the changes in tidal flushing due to sea-level fluctuations, it is the position and height of the sea in relation to the land [Relative Sea Level (RSL)] that is important, as this determines the location of the shoreline. In addition to global fluctuations in sea levels (resulting from the growth and melting of continental glaciers and thermal expansion of oceanic waters), many physical processes may result in changes of the RSL at local-to-regional scales. These include changes in meltwater load, crustal rebound from glaciation, uplift or subsidence in coastal areas related to various tectonic processes, fluid withdrawal, and sediment deposition and compaction (see "RSL in the Venice lagoon: the role of subsidence"). Moreover, the water mass structure and the water fluxes and exchanges with adjacent seas are concurrent with long-term sea-level fluctuations. In particular, most of the Mediterranean sub-basins have an evaporative nature, resulting from strong evaporation and weak river runoff, so that in recent years their sea-level trends have been substantially different from those based on global ocean-related estimates (see "RSL in the Venice lagoon: the role of seasonal to decadal climatic fluctuations").

The tide gauges located in the Venice lagoon and inside the city provide a direct measurement of the RSL, as they incorporate landmass movements (natural and urban subsidence, the subsidence of the man-made structures on which tide gauges are mounted and changes in the seabed and coastal topography), which are read by the instruments as sea-level variations. Thus, even if gauge records are coherent and indicative, they are difficult to interpret in dynamic and evolutionary terms. In fact, although the direct measurement of the parameter "perceived" by the coast is the best value of gauge measurements, this also represents their major limitation as they cannot lead to a reliable sea-level change estimate unless independent estimations of local crustal movements are available. Hence, the necessity for estimating individual contributions to the RSL is crucial in order to provide reliable future sea-level fluctuation scenarios.

Once the possible ecological impacts of fluctuations in the RSL in the Venice lagoon have been presented (see "Sea level rise in the Venice lagoon: ecological impacts") and the contribution of local subsidence (natural and anthropogenic) in the Venice lagoon have been estimated (see "RSL in the Venice lagoon: the role of subsidence"), the fraction of the RSL fluctuations measured in the lagoon and attributable to effective sea-level changes is calculated and linked to climatic fluctuations on a regional scale, through the hydrodynamics induced in the Northern Adriatic by the freshwater inflow from the Po River (see "RSL in the Venice lagoon: the role of seasonal to decadal climatic fluctuations"). Concluding remarks follow, with some criticisms of former forecasts for next century sea-level rise in the Venice lagoon.

Sea level rise in the Venice lagoon: ecological impacts

The bed of a lagoon is the substrate upon which biodiversity is dependent. Changes in the morphology can induce a shift in the biological communities that reside in a defined area by changing the potential vocation of the area itself. Given the great variability in biomass and biodiversity which characterizes the Venice lagoon (i.e. low energy and highly confined zones with low biodiversity; marshlands and gutters; canals; protected marine environments with high biodiversity; areas near the mouths with strong hydrodynamics and tidal renewals), it is fundamental to assess the areas where the ecosystem functionality may be compromised. In fact, over recent decades the lagoon has been suffering from severe environmental degradation, which was induced also by complex morphodynamic changes caused by natural processes (geomorphologic variations have been naturally occurring in the lagoon since its formation) and by the direct and indirect impact of human activities. As early as the seventies (Montanelli et al., 1970) there were great concerns about the loss of a large proportion of tidal flat habitats in the lagoon—up to 160 km^2 on the northern, southern and inner borders of the lagoon—because of human interventions (amongst others: the construction of the Malamocco canal, which services the industrial area of Marghera, and fish farms). The ecological importance of these habitats is not only naturalistic, because of the biological communities they host, but also environmental, through their role in controlling the water quality by sequestrating the industrial and domestic pollutants that are discharged into the lagoon. This is the reason why tidal flushing is central to the well-being of the lagoon, and many scientific and technical interventions currently focus on controlling and recovering the tidal flats in an attempt to limit their natural erosion (from the effect of waves and wave resuspension).

Despite the great morphodynamic complexity of the tidal flats (currently the northern part of the lagoon is witnessing a tendency for the growth of tidal flats while the southern and central parts have been scoured and deepened—see Amos et al., 2002), which includes a link between marsh elevation, sea-level changes and sediment deposition/resuspension (Reed, 1995), the flows through the three mouths and the materials carried by the main tidal channels are likely to be key factors in balancing the sediment budget, and in turn the geomorphological evolution of marshes.

Thus, particular emphasis should be placed on comparing sedimentation rates and contemporary RSL estimates. With a scenario of rising sea-levels and given the micro- to meso-tidal nature of the subsiding lagoon (see "RSL in the Venice lagoon: the role of subsidence"), a sedimentation deficit is a likely consequence of increased erosion due to higher energy waves (Stevenson et al., 1986). Because of the inadequate sediment supply, the elevations of marsh systems may not adjust to even a moderate rate of sea-level rise, which could lead to a substantial reduction of the tidal flats. The ecological effects are likely to be significant for the benthic assemblages of tidal flats (for a review see Raffaelli and Hawkins, 1996) and for the consumers they support, especially fish, shrimps and shorebirds, through complex and not easily predictable biological mechanisms. In addition to the loss of intertidal or shallow sublittoral areas, a rise in sea level would provide more habitats for salt tolerant flora and fauna, and fewer habitats for freshwater marsh grasses, thus changing nutrient and phytoplankton distributions in the lagoon.

Therefore, an understanding of both (a) the hydrodynamics in the Venice lagoon and of the erosion-transport-sedimentation processes, as well as their interrelations, and (b) soil subsidence and sea-level rise, are fundamental for preserving the delicate lagoon ecosystems. Great effort has already been put into understanding the phenomena of subsidence, sediment erosion, re-suspension, transport and sedimentation, sea-lagoon balance and hydrodynamics, as well as to the planning and management of interventions and protection works: several investigations have been conducted, including geomorphological and sedimentological studies (Ciavola et al., 2002; Rizzetto et al., 2003, respectively), profiles (Cola & Simonini, 2002), and mapping (Strozzi et al., 2002); a number of mathematical models have also been developed, focussing on hydrodynamics and morphology (Umgiesser, 2000; Bergamasco et al., 2001; Amos et al., 2002; Umgiesser et al., 2002; Bonardi et al., 2003). In this connection, once reliable estimates of interannual to decadal RSL fluctuations have been defined, these models may become optimal tools for evaluating how the distribution of the biologic communities in the lagoon would change in relation to sea-level changes, through distribution gradients related to hydro-morphological and chemical-physical conditions.

RSL in the Venice lagoon: the role of subsidence

The Venice lagoon's increased vulnerability to sea-level fluctuations is partly due to its location in an active area that is naturally subject to both tectonic and sedimentological processes, and partly to the consequences of human activities (i.e. urbanism, artesian withdrawals and modifications to the lagoon morphology).

The Po Plain encompasses an area of about 38,000 km^2 south of the Alps mountain range and its geodynamic features are related to long-term processes (plates subduction and Quaternary sediments compaction), and to short-term processes [PostGlacial Rebound (PGR) and differential compaction]. The PGR, which is the slow readjustment and rebalancing of the lithosphere and mantle during deglaciation after the downflexure caused by the Alpine ice cap during the last Ice Age, controls approximately 50% of the subsidence rates in the Po Plain (which range between 0 and 5 mm/yr) (Carminati et al., 2003b). It has been inducing temporal and spatial variability in the landmass movements since the Late Quaternary (different subsidence rates in the areas close to and far from the Alps) and they are still likely to be active (Mitrovica & Davis, 1995). Moreover, the Venice lagoon is located on a segment of the active southwest dipping monocline that is related to subduction in the northern Apennines, so that a significant part of the natural component of lagoon subsidence could be related to this downflexure (Carminati et al., 2003a). The solid material that is transported by rivers flowing into the Adriatic sea is distributed along the coast by the currents and this counterbalances the coastal lowering rate, while the subsidence effect in the Venice lagoon is not offset by sediment accumulation since its most important tributaries have been artificially diverted away from the lagoon since the XV century.

Subsidence rates within the Venice lagoon are spatially differentiated (the levelling campaigns of 1973 and 1993 showed rates ranging between 1.4 and –0.5 mm/yr: Teatini et al., 1995). In the second half of the 20th century, this variability was mainly affected by human activity, particularly urban subsidence and groundwater extraction. A comparative analysis of artesian exploitation and subsidence identified three distinct periods (Carbognin et al., 1977): the first before 1952, when artesian exploitation was not very intensive and subsidence was only due to natural causes; the second from 1952 to 1969, when artesian water extraction was very active, causing a local average subsidence rate in the city of over 9 cm, with local maxima of 10 cm; the last period, after water extraction had been stopped, was characterized by a period of stability and a subsequent ground-surface rebound (more than +2 cm in the historic centre in 1975 compared to 1969). At present, land subsidence maps of the lagoon (Strozzi et al., 2002) and the consistence between tidal records in Venice, Trieste, Rovinj and Bakar indicate that subsidence triggered by human activity is no longer an issue for Venice (apart from recently urbanized areas such as the isle of St. Elena, where sediment compaction is active), and that the ground level in the historical centre of Venice is almost stable and only subject to natural subsidence.

Hence it is now crucial to assess the natural subsidence rates in the area, their origin and variations over time. In particular, the difference between the estimates of long-term (10^6 yr) natural subsidence affected by the subduction associated with the Apennines (0.7–1.0 mm/yr, see Carminati et al., 2003a) and present-day natural subsidence in the lagoon (approximately 0.5 mm/yr, see Tosi et al., 2002) would suggest that the natural short-term component (10^3–10^4 yr) is still dominant in the Venetian area, since it is likely to be related to climatic changes such as the PGR. Thus, although the PGR effect has been diminishing, significant variations in the natural subsidence rate are not likely to occur in the Venice lagoon within the next century. This would mean that a lowering of the ground level in the order of 5 cm from present levels would seem to be a realistic estimate for the year 2100.

RSL in the Venice lagoon: the role of seasonal to decadal climatic fluctuations

The inter-relation between the mean sea level, the resulting circulation and wave climates in the Northern Adriatic may induce changes in the sea level inside the Venice lagoon, both in terms of tidal extremes and of seasonal to decadal fluctuations. In particular, the evolution of the meteorological and regional climatic parameters induce hydrostatic responses (through the 'inverse barometer' effect, see Anthes, 1982) and non-hydrostatic responses (thermoaline circulation and wind-driven Ekman Layer transport) in the sea levels of the Northern Adriatic. Notably, sea-level fluctuations occurring in the Mediterranean Sea and in its sub-basins on interannual to decadal scales are generally much larger than those associated with secular trends (Woolf et al., 2003), because of their evaporative nature which makes them sensitive to changes in precipitation/evaporation budgets. Changes in decadal sea-level trends in the Mediterranean are likely to be induced by transitions in the processes of deep water formation, through changes in the temperature and salinity of deep and intermediate layers (Tsimplis & Baker, 2000). Moreover, the Adriatic Sea's shallow northern part, with an average depth of about 30 m, has strong continental characteristics so that its hydrodynamics and oceanographic parameters are highly dependent on the freshwater inflow, mainly through the induction of density currents. The plume of the River Po, whose formation and evolution are linked to the river's flow rates, seawater stratification and the wind stress from the Bora and Scirocco, adds to the complexity of the baroclinic geostrophic structure of the general circulation in the basin. Under typical winter conditions (November to March), when a cyclonic gyre north of the Po River delta—Rovinj line is present (Krajcar, 2003), the plume of the Po is confined to the western Adriatic shelf region. Nevertheless, following prolonged intense Bora events, the plume appears to be entrained into the circulation of the northern cyclonic gyre induced by the wind stress curl applied by the Bora (Paklar et al., 2001). Upwelling-favourable winds such as Scirocco reverse the western coastal current and advect Po waters towards the northern coast and offshore (Alberotanza et al., 2004).

In order to estimate the relationships between the Po River discharges and sea levels in the Venice lagoon, the monthly-mean values of the 'Punta della Salute' dataset (the gauge is sited in the historical centre of Venice) and the monthly-mean Po River discharges at Pontelagoscuro for the period 1968–2001 (after the worst flooding event ever recorded in Venice, occurred in the autumn of 1966) have been considered. The sea-level dataset was provided by the 'Ufficio Idrografico' (Hydrographic Office) of Venice and has been conventionally chosen as the reference data for the RSL in the Venice lagoon. The harmonic components of astronomical tides and the contribution of subsidence were subtracted from the observed sea-level, in order to obtain the meteorological contribution. Subsidence was estimated in accordance with the model by Carbognin et al. (1977), assuming the present-day subsidence rate to be 0.5 mm/yr.

The linear correlation ($r = 0.58$ for 34 cases, significant at $p < 0.001$) between the October–March averages of discharge and sea level (Fig. 2) would suggest the existence of a relationship; however, nothing can be deduced about the phase relationships or the causality between the two variables.

In order to obtain further information, which was not readily available from the raw dataset, a Wavelet Analysis was conducted on the signals to identify time/frequency localization. Wavelet Analysis, some background of which is provided in the paper by Torrence & Compo (1998), is based on mathematical transformations that decompose a one-dimensional time series into a diffused two-dimensional time-frequency image simultaneously, providing a means for getting information on both the amplitude (and the phase) of any 'periodic' signal in a time series, and how this amplitude varies with time. This information is obtained through the projection of a generic function (a wave function with a specific frequency and finite duration) on the data, by sliding the wavelet along the time series, and scaling it by changing its width. The generic function is called the *mother wavelet* $\phi(x)$ (in this study the Morlet function was used), defined as:

$$\phi(x) = \sum_{k \in Z} c_k \phi(2x - k) \quad (1)$$

where c_k is a set of real coefficients that verify regularity, orthogonality and normalization requirements. All the elementary basis functions are obtained by translating and scaling the mother wavelet through a *scaling function* $\Psi(x)$, defined as:

$$\Psi(x) = \sum_{k \in Z} (-1)^k c_{1-k} \phi(2x - k) \quad (2)$$

The scaling is the main advantage of the Wavelet Analysis over a Windowed Fourier Transform, whose basic elements are sines and cosines, which solves the frequency localization problem but is dependent on the window size used, so that different frequencies are treated inconsistently. The Wavelet Power Spectrum (WPS), defined as the power (absolute value squared) of the wavelet transform of the original series, provides information on the oscillations within the signal at a certain scale and a certain time. Significance levels are established by comparing the wavelet spectrum of the variable under study with background theoretical spectra of a red-noise (univariate lag-1 autoregressive) process.

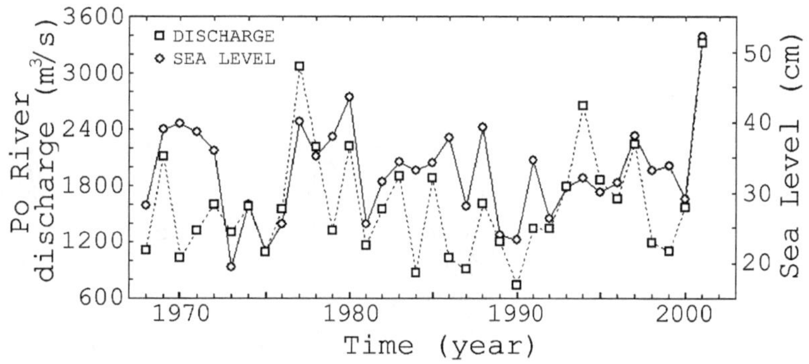

Fig. 2 October–March averages of Po River discharges and of sea-levels measured in the Venice lagoon

The WPS of discharges and sea levels shown in Fig. 3 and 4 respectively, indicate that there is a greater concentration of significant power in bands below the 16-month period. The significant peaks within the 4–8-month band in Fig. 3 (around 1968, 1977, 1983–1984, 1994 and 2000–2001), which are indices of high-power seasonal signals, are concurrent to high discharges from the Po River, and not necessarily linked to longer wet periods. Interestingly, groups of peaks at these frequencies are separated by periods of low power with a mean length of about 8 years. It is worth noting that the largest significant peaks in the 4–8 months band in the WPS of sea-levels appear in periods when the seasonal signal of the discharge is low (around 1991–1993 and 1997). Within the 8–16-month band (annual), significant peaks in sea levels (around 1970–1971, 1976–1982, 1992–1994 and 1997) are much larger than in shorter frequencies; the opposite behaviour emerges from the spectrum of the discharges (around 1972, 1977–1978, 1986). It is also worth pointing out that in the annual band the significant peaks of the two time series alternate, with the exception of the concurrent peaks around 1977–1978.

The characteristics of a WPS are reflected in the Global Wavelet Spectrum (GWS), which is obtained by averaging in time the WPS over all the local wavelet spectra and provides a measure of the variation of energies across scales. The GWS in Fig. 5 shows the dominating frequencies to have period of 6 months (discharges and sea levels) and 12 months (discharges). Notably, despite the GWS of discharges and sea levels not being significant at low frequencies, both present an important peak around the 8-year period (8.25 and 7.35 years, respectively). Significance levels were derived assuming red-noise spectra with lag-1 autocorrelation parameters $\alpha = 0.49$ for discharges and $\alpha = 0.34$ for sea levels.

The normalized Wavelet Coherence (Maraun & Kurths, 2004) is a bivariate extension of the Wavelet Analysis, defined as the expectation value of the product of the two corresponding wavelet transforms (which defines the Wavelet Cross Spectrum), normalized to the two single WPS. It individuates regions with large common power in the time-frequency domain of two time series and further reveals information about their phase relationship: if two series are physically related (which may be suggestive of causality) a consistent or slowly varying phase lag is expected and the circular mean of the phase angles can be used to quantify the phase relationship. The Wavelet Coherence spectrum of discharges and sea levels is reported in Fig. 6. Although the 5% significance level is not a reliable indication of causality, the significant region of the spectrum is so extensive that it seems very unlikely that this is simply a chance occurrence, at least on seasonal (around the 6-month period) and interannual (periods of 16 to 32 months) wavelengths.

Discussion

Long-term cyclic variations in sea level in the Mediterranean Sea are partly due to the interannual/decadal variability of the upper ocean circulation (Cazenave et al., 2001). Nevertheless, sea-level fluctuations induced by changes in deep water formation processes and by hydrostatic and non-hydrostatic responses to the climate variabil-

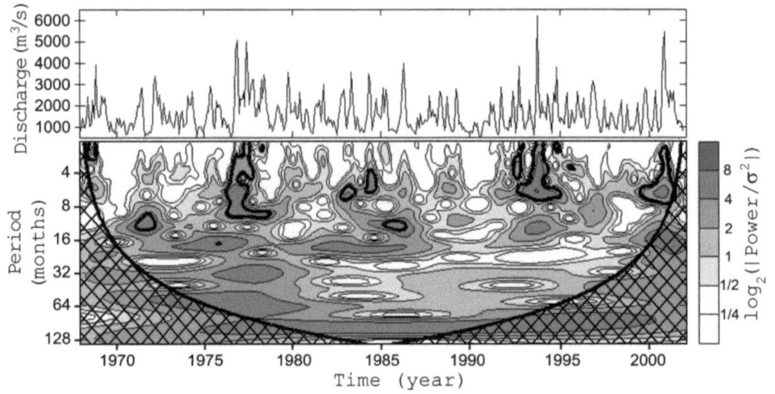

Fig. 3 Time series of monthly discharges of the Po River for the period 1968–2001 (top) and corresponding Wavelet Power Spectrum (bottom). The power is normalized by $1/\sigma^2$. The black thick contour is the 5% significance level for a red-noise AR(1) process with lag-1 of 0.49. The cross-hatched region is the cone of influence, where edge effects occur

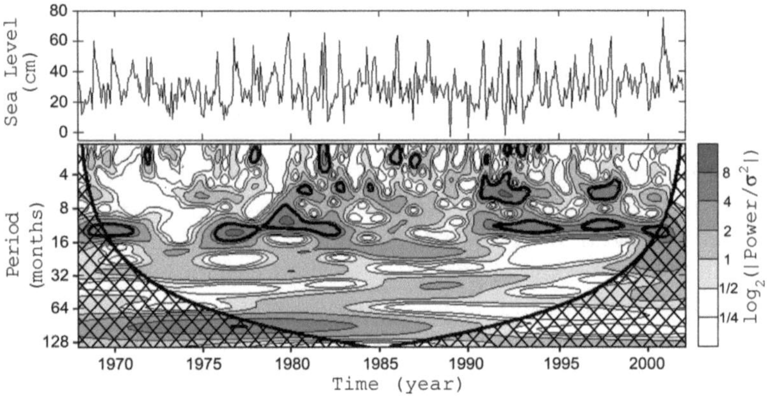

Fig. 4 Time series of monthly sea levels in the Venice lagoon for the period 1968–2001 (top) and corresponding Wavelet Power Spectrum (bottom). The power is normalized by $1/\sigma^2$. The black thick contour is the 5% significance level for a red-noise AR(1) process with lag-1 of 0.34. The cross-hatched region is the cone of influence, where edge effects occur

ity of the region, make it impossible to attain forecasts based on estimates of the global sea-level rise alone. The sea-level increase throughout the Eastern Mediterranean area during the 1990s, which was comparable to the rise in global sea-level rates recorded by the Intergovernmental Panel on Climate Change (IPCC), was concomitant to a prolonged positive phase of the North Atlantic Oscillation (NAO) (Tsimplis & Baker, 2000), which is the foremost mode of climate variability in the North Atlantic region (Hurrell, 1995) as it controls the synoptic weather over the entire Euro-Mediterranean area. Remarkably, satellite altimeter measurements (Topex/Poseidon, 1992–2001) indicate the Northern Adriatic basin as one of the areas that is most sensitive to fluctuations in the NAO, despite its influence in the Mediterranean Sea being quite homogeneous (Woolf et al., 2003).

Thus, coastal zone management would benefit by the coupling of long-term future projections of sea levels (as those based on global sea-level rise rates) with indications of the expected local interannual-to-decadal fluctuations (as those driven by the NAO). All the more so because in the Mediteranean decadal sea-level fluctuations are

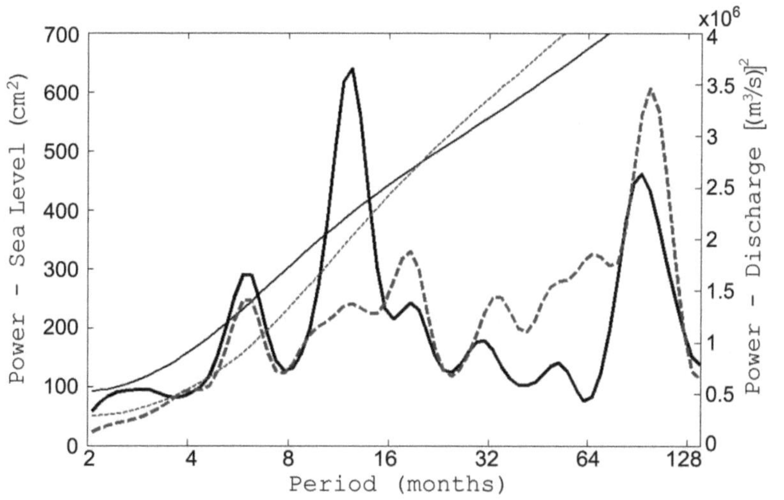

Fig. 5 Global wavelet spectra of monthly Po River discharges (black line) and sea levels in the Venice lagoon (dotted line) for the period 1968–2001, with the 5% significance levels calculated from backward red-noise spectra

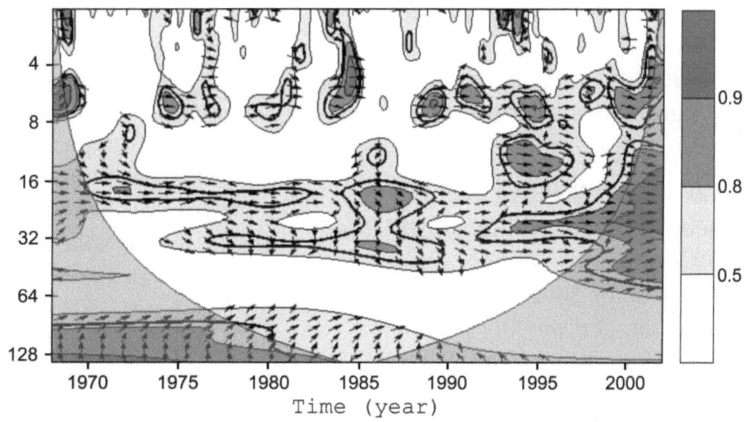

Fig. 6 Wavelet coherence between monthly Po River discharges and sea levels in the Venice lagoon for the period 1968–2001. The 5% significance level against red noise, tested with a Monte Carlo method (10,000 surrogate datasets), is shown as a thick contour. The relative phase relationship is shown as arrows (with in-phase pointing right, anti-phase pointing left and discharge leading sea levels by 90° pointing straight down). The shaded region is the cone of influence, where edge effects occur

far more important than secular trends (Woolf et al., 2003). In fact, the estimates for the future sea-level rise in the Venice lagoon in the year 2100 obtained from former studies from the IPCC (47 ± 39 cm/cy, see Church et al., 2001) and from the Consortium for Research in the Venice lagoon (CO.RI.LA., 1999) (a 'most probable' scenario of +16.4 cm/cy, of which 12.3 cm/cy due to sea-level changes) cannot really provide an effective picture of the nearest 'most probable' future.

In this context, the individuation of phase-locked fluctuations on interannual to decadal scales between sea levels and predictable climatic parameters may suggest sea-level patterns over the next decades. In particular, discharges are a function of the spatial integration of precipitation over the river's catchment. Besides the hydrological processes linking rainfalls to discharge are affected by non-linear processes (such as infiltration, evapotranspiration and snow melt) and various human interventions (such as channel and catchment changes, and management of water reservoirs), the analysis of discharge records provides an immediate insight into climatic conditions. In the case of the Po River, the discharges measured at Pontelagoscuro (the closing section of the river, located about 90 km upstream the Po Delta) are indicative of the precipitation variability over the whole Po catchment, which has an extension of about 70,000 km^2, hence providing indications of the regional climate. As discussed in "RSL in the Venice lagoon: the role of seasonal to decadal climatic fluctuations", monthly Po River discharges and sea levels in the Venice lagoon show wavelet spectra with peculiar but related characteristics, above all on seasonal to interannual scales: (a) periods with significant peaks alternate with low power periods; (b) opposite behaviour of high-frequency power signals emerges from the two spectra; (c) global power spectra has peaks at similar periods; (d) phase relationships emerge between the two time series.

Further observations can be drawn from these results.

(1) Interestingly, peaks in the WPS of the Po River discharges seem to occur within active phases of the NAO, which are not necessarily positive as in the case of 2000–2001. The patterns of the scale averaged wavelet power (defined as the weighted sum of the WPS over a defined interval of scales) over the 8–16 month band for the Po River discharges, and of the smoothed absolute values of the NAO index in Fig. 7, are likely to confirm the existence of a link between dominant zonal or meridional atmospheric circulation (controlled by the NAO) and seasonal to interannual precipitation in regions south of the Alps, an observation which has already been suggested by Quadrelli et al. (2001).

(2) Phase relationships shown in the Wavelet Coherence spectrum (Fig. 6) indicate that the oscillatory patterns of discharges and sea levels are dynamically synchronized across a wide region of temporal and frequency domains: they are in-phase in the 4–8 month band, as a consequence of their similar seasonal patterns, and discharges generally lead sea-level fluctuations on an interannual scale. Non-phase-locked periods, such as the 1980s, are a consequence of the non-linear responses of regional climatic parameters to the sources of their variability. In this context, it is worth remembering that the 1980s were a particularly active period for the climate on a global scale: they followed the 'climate shift' of 1977 (Hare & Mantua, 2000), were affected by the atmospheric effects of the 1982 volcanic eruption of the El Chichon (Mexico) and by two El Niño events (a strong one in 1982–1983 and a weaker one in 1986–1987), as well as a strong La Niña event in 1988–1989 (Philander, 1990). Further discussions about the response of Po River discharges and derived indices of drought to climatic fluctuations and extreme events occurring on a regional scale can be found in Tomasino et al. (2004b).

(3) Finally, the presence of peaks around the 8-year period in the GWS of both time series (Fig. 5), although not significant, may be related to the climatic variability induced by solar forcing, whose frequency spectrum is linked to the solar wind and has one of its strongest peaks at the 8.6-year period (Landscheidt, 2000). In this connection, a study on the feasibility of long-term forecasting of seasonal discharges of the Po River based on calculable solar is available in Tomasino et al. (2004a).

Conclusions

Variations in sea-level height, from the diurnal oscillations to the decadal and secular fluctuations forced by climate changes, are the dominant forces in the dynamics of the Venice lagoon ecosystems. An understanding of sea level variability within the lagoon and in the adjacent basin (the Northern Adriatic) and of the forces that induce their fluctuations, is therefore necessary in order to gain a clearer understanding of how the delicate natural equilibriums of the ecotone can be preserved in the future. Short-term phenom-

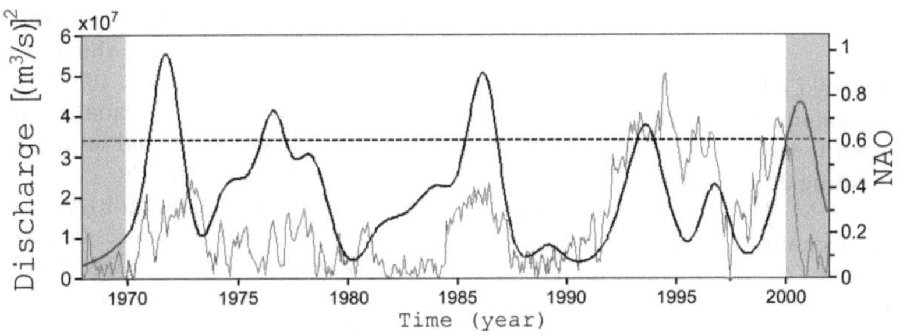

Fig. 7 Time averaged wavelet spectrum of the monthly Po River discharges for the 8–16-month scales (black line) with 5% significance level (dotted line), and absolute values of the 3-year centred smoothing of the monthly NAO index (grey line). Shaded area indicates where edge effects occur

ena such as storm surges, which are a consequence of peculiar meteorological conditions coupled with unfavourable astronomical phases, can seriously jeopardize the whole lagoon ecosystem. Nevertheless, sea-level fluctuations on longer time scales (interannual to decadal) may be as destructive for the ecotone as high-energy events can be, since they slowly alter the hydrodynamics, and in turn modify the active/passive areas of the lagoon, the distribution of nutrients and biological communities, may favour or prevent sediment deposition/resuspension and eutrophization, as well as alter the actual vocation of special areas such as tidal flats.

Assuming that the interannual and decadal variability of sea levels in the Adriatic Sea is inextricably linked to the Atlantic Sector response, the dominant influence of the NAO, coupled with the peculiar nature of the Northern Adriatic basin, are key factors for gaining an accurate understanding of the long-term dynamics of sea levels in the Venice lagoon. In fact, the main limitation of former projections of sea-level changes in the Venice lagoon for the next century was that they neglected the latter in favour of estimating rises in global sea-level rates (by IPCC and CO.RI.LA.). Moreover a secular rate is used in these projections to predict the sea-level height in the year 2100, while the interannual to decadal variability is much larger than the secular trend and may in fact obscure it, at least in the upcoming decades.

In this preliminary study, the seasonal to interannual sea-level variability in the Venice lagoon, its hydrological connections with the Northern Adriatic and specifically with the Po River, which is the major contributor of freshwater runoff in the basin, were analyzed through Wavelet Analysis techniques. In particular, this study focussed on the phase relationships between sea levels in the Venice lagoon and Po River discharges, assessing that oscillatory patterns appear to be dynamically synchronized across a wide region of the temporal and frequency domains, i.e. on seasonal to interannual scales. The coupled dynamics of discharges and sea levels would suggest a response by the regional climate system of Northern Italy/Northern Adriatic to large scale climatic patterns, i.e. the North Atlantic sector whose variability is dominated by the North Atlantic Oscillation, and also to global climate shifts and hazardous events, such as El Niño events or explosive volcanic eruptions.

The results presented, which contribute to mitigate some daring statements about what the Venice lagoon is facing, may be used as basic indications for further studies, included the development of predictive tools at the seasonal to interannual scales, that could provide a reliable means for finding an answer to some difficult questions about the future of Venice

Acknowledgements The authors would like to thank Prof. A. Tomasin from the University of Venice and P. Canestrelli from the 'Centro Previsioni e Segnalazioni Maree' for kindly providing the Venice sea-level data, and Dr. Carlo Allodi for kindly providing the discharge data about the Po River. The authors also thank two anonymous reviewers for help in revising this manuscript.

References

Alberotanza, L., J. Chiggiato & G. Profeti, 2004. Analysis of the seasonal stratification at the Acqua Alta Oceanographic Tower, northern Adriatic Sea. International Journal of Remote Sensing 25(7–8): 1473–1480.

Amos, C. L., G. Umgiesser, M. Bonardi & S. Cappucci, 2002. Sedimentation mechanisms in northern Venice lagoon. In Campostrini, P. (ed.), Proceedings of IVSLA-Corila Research Program 2001 results, 315–332 (ISBN 88–88143–12–2).

Anthes, R. A., 1982. Tropical Cyclones, Their Evolution Structure and Effects. American Meteorological Society, Boston, Mass.

Bergamasco, A., S. Carniel & G. Umgiesser, 2001. A simple hydrodynamic-deposition couplet model of the lagoon of Venice. Biologia Marina Mediterranea 8(1): 300–308.

Bonardi, M., L. Tosi, A. Cucco, G. Umgiesser, & F. Rizzetto, 2003. Geomorphic processes and hydrodynamics in the Venice Lagoon, Italy: a case study. Hydrology of the Mediterranean and Semiarid Regions. In Proceedings of an International Symposium held at Montpellier, April 2003, IAHS Publ. no. 278, 448–457.

Canestrelli, P., M. Mandich, P. A. Pirazzoli & A. Tomasin, 2001. Winds, Depressions and Seiches: Tidal Perturbations in Venice (1951–2000) (bilingual). Centro Previsioni e Segnalazioni Maree, Venezia, 105 pp.

Carbognin, L., P. Gatto, G. Mozzi & G. Gambolati, 1977. Guidebook to Studies of Land Subsidence due to Ground Water Withdrawal. IAHS-AISH Publication No. 121, 65–81.

Carminati, E., C. Doglioni & D. Scrocca, 2003a. Apennines subduction-related subsidence of Venice (Italy). Geophysical Research Letters 30(13): 1717–1723.

Carminati, E., G. Martinelli & P. Severi, 2003b. Influence of glacial cycles and tectonics on natural subsidence in the Po Plain (Northern Italy): Insights from ^{14}C ages. Geochemistry Geophysics Geosystems 4(10), ISSN: 1525-2027.

Cazenave, A., C. Cabanes, K. Dominh & S. Mangiarotti, 2001. Recent sea level change in the Mediterranean sea revealed by Topex/Poseidon satellite altimetry. Geophysical Research Letter 28(8): 1607–1610.

Church, J. A., J. M. Gregory, P. Huybrechts, M. Kuhn, K. Lambeck, M. T., Nhuan, P. Qin & P. L. Woodworth, 2001. Changes in sea level. In Climate Change: the Scientific Basis. Cambridge University Press, New York: 501–555.

Ciavola, P., C. Organo, L. León Vintró & P. I. Mitchell, 2002. Sedimentation processes on intertidal areas of the lagoon of Venice: identification of exceptional flood events (Acqua Alta) using Radionuclides. Journal of Coastal Research SI 36: 139–147.

CO.RI. LA., 1999. Scenari di crescita del livello del mare per la Laguna di Venezia. Co.Ri.La. 1: 1–40.

Cola, S. & P. Simonini, 2002. Mechanical behaviour of Venice lagoon silts as a function of their gradient characteristics. Canadian Geotechnical Journal 39(4): 879–893.

Hare, S. R. & N. J. Mantua, 2000. Empirical evidence for North Pacific regime shifts in 1977 and 1989. Progress in Oceanography 47: 103–145.

Hurrell, J. W., 1995. Decadal trends in the North Atlantic Oscillation: Regional temperatures and precipitation. Science 269: 676–679.

Krajcar, V., 2003. Climatology and geostrophic currents in the Northern Adriatic. Geofizika 20: 105–114.

Landscheidt, T., 2000. River Po discharges and cycles of solar activity. Hydrological Science Journal 45(3): 491–493.

Maraun, D. & J. Kurths, 2004. Cross wavelet analysis: significance testing and pitfalls. Nonlinear Processes in Geophysics 11: 505–514.

Mitrovica, J. X. & J. Davis, 1995. Present-day post glacial sea level change far from the Late Pleistocene ice sheets: Implications for recent analyses of tide gauge records. Geophysics Research Letter 22: 2529–2532.

Montanelli, I., G. Samona & F. Valcanover, 1970. Venezia, Caduta e Salvezza (Venice, Fall and Salvation; in italian). Publ. Sansoni Nuovi Spa, 28 pp.

Paklar, G. B., V. Isakov, D. Koracin, V. Kourafalou & M. Orlic, 2001. A case study of bora-driven flow and density changes on the Adriatic Shelf (January 1987). Continental Shelf Research 21: 1751–1783.

Philander, S. G. H., 1990. El Niño, La Niña and the Southern Oscillation. Academic Press, San Diego, CA, 289 pp.

Quadrelli, R., M. Lazzeri, C. Cacciamani & S. Tibaldi, 2001. Observed winter Alpine precipitation variability and links with large scale circulation patterns. Climate Research 17(3): 275–284.

Raffaelli, D. & S. J. Hawkins, 1996. Intertidal Ecology. Chapman and Hall, London.

Ravera, O., 2000. The lagoon of Venice: the result of both natural factors and human influence. Journal of Limnology 59(1): 19–30.

Reed, D. J., 1995. The response of coastal marshes to sea-level rise: survival or submergence? Earth Surface Processes and Landforms 20: 39–48.

Rizzetto, F., L. Tosi, L. Carbognin, M. Bonari & P. Teatini, 2003. Geomorphological setting and related hydrogeological implications of the coastal plain south of the Venice Lagoon (Italy). In E. Servat et al. (ed.), Hydrology of the Mediterranean and Semiarid Regions, IAHS Publ. n. 278, 463–470.

Stevenson, J. C., L. G. Ward, & M. S. Kearney, 1986. Vertical accretion in marshes with varying rates of sea level rise. In Wolfe D. A. (ed.), Estuarine Variability. Academic Press, Orlando, Florida, 241–259.

Strozzi, T., L. Tosi, U. Wegmueller, P. Teatini, L. Carbognin & R. Rosselli, 2002. Thematic and land subsidence maps of the lagoon of Venice from ERS SAR interferometry. In Campostrini, P. (ed.), Proceedings of IVSLA-Corila Research Program 2001 results, 345–355 (ISBN 88-88143-12-2).

Teatini, P., G. Gambolati & L. Tosi, 1995. A new 3-D nonlinear model of the subsidence of Venice. In Proceedings of the 5th International Symposium on Land Subsidence, IHAS Publication n. 234, 353–361.

Tomasino, M., D. Zanchettin & P. Traverso, 2004a. Long-range forecasts of River Po discharges based on predictable solar activity and a fuzzy neural network model. Hydrological Sciences—Journal—des Sciences Hydrologiques 49(4): 641–652.

Tomasino, M., D. Zanchettin & P. Traverso, 2004b. Analisi dell'evento siccitoso dell'estate 2003 con riferimento alla serie storica (1807–2002) dei dati di portata del Po a Pontelagoscuro (Analysis of the drought of summer 2003 based on the time series (1807–2002) of the Po River discharges at Pontelagoscuro). Bollettino Geofisico n.1–2: 11–22.

Torrence, C. & G. P. Compo, 1998. A practical guide to wavelet analysis, Bulletin of American Meteorological Society 79: 61–78.

Tosi, L., L. Carbognin, T. Teatini, P. Strozzi & U. Wegmüller, 2002. Evidence of the present relative land stability of Venice, Italy, from land, sea, and space observations, Geophysics Research Letter 29(12), 1562, doi: 10.1029/2001GL013211.

Tsimplis, M. N. & T. F. Baker, 2000. Sea level drop in the Mediterranean Sea: An indicator of deep water salinity and temperature changes? Geophysics Research Letter 27(12): 1731–1734.

Umgiesser, G., 2000. Modelling residual currents in the Venice Lagoon. In Yanagi, T. (ed.), Interactions between Estuaries, Coastal Seas abd Shelf Seas, TER-RAPUB, 107–124.

Umgiesser, G., M. Sclavo & S. Carniel, 2002. Modelling the bottom stress distribution in the Venice lagoon. In Campostrini, P. (ed.), Proceedings of IVSLA-Corila Research Program 2001 results, 287–299 (ISBN 88-88143-12-2).

Woolf, D., A. Shaw & M. Tsimplis, 2003. The influence of the North Atlantic Oscillation on sea level variability in the North Atlantic Region. Global Atmosphere and Ocean System 9(4): 145–167.

Hydrobiologia (2007) 577:55–70
DOI 10.1007/s10750-006-0417-4

LAGOONS AND COASTAL WETLANDS

Defining the coast and sentinel ecosystems for coastal observations of global change

Robert R. Christian · Stefano Mazzilli

Abstract The detection, attribution and prediction of global and large scale regional change are goals for the Global Observing Systems of the United Nations. Coastal areas are particularly sensitive to global change, but there is a variety of limitations to universal coverage of observations. The coastal module of the Global Terrestrial Observing System (C-GTOS) considers sentinel ecosystems to address these goals for the terrestrial, wetland and freshwater ecosystems of the coast. Sentinel ecosystems for observing systems are a limited number of well understood systems that have substantial datasets and are observed in a sustained fashion, forming an early warning and core system for broader regional and global change. A necessary step in the development of C-GTOS is the examination of current definitions of coastal areas by anticipated users and information providers, and identification of potential coastal networks and sites. We applied the sentinel system framework to the selection of C-GTOS observation sites from several international programs using various global delineations of coastal areas. Delineations were based on the most common definitions of the coast adopted by potential C-GTOS users and information providers, and included mapped areas of various distance from the coastline, coastal areas of low elevation, and a seaward boundary matching the Economic Exclusive Zone (EEZ). Decreases in the number of sites within each international program occurred with each definition marking area closer to the coastline. The Ramsar Convention on Wetlands demonstrates the greatest percentage of coastal sites by any definition. The process of choosing specific sentinel sites for C-GTOS continues from this initial screening, and is the next step towards the development of an in situ site network supporting the observation of global and large scale change.

Keywords Coast · Global Terrestrial Observing System · ecosystems · Long-term monitoring · Ramsar Convention on Wetlands · World Heritage Convention

Guest editors: P. Viaroli, P. Lasserre & P. Campostrini
Lagoons and Coastal Wetlands in the Global Change Context: Impacts and Management Issues

R. R. Christian (✉)
Global Terrestrial Observing System/Biology Department, East Carolina University, Greenville, NC 27858, USA
e-mail: CHRISTIANR@ECU.EDU

S. Mazzilli
Global Terrestrial Observing System/Environment and Natural Resources Service, Food and Agricultural Organization of the United Nations, Rome 00185, Italy

Introduction

The detection, attribution and prediction of global and large scale regional change have become goals for numerous national and multinational organizations. The United Nations contributes to these goals through its Global Observing Systems. The observing systems are divided into three major programs: Global Terrestrial Observing System (GTOS), Global Ocean Observing System (GOOS), and Global Climate Observing System (GCOS). The coast is one area where global change appears critically important and where interaction among the observing systems is essential for effectively meeting their goals. Both GOOS and GTOS are developing coordinated coastal programs and contributing to the Integrated Global Observing Strategy (IGOS), the observing system strategy fostering cooperation among various UN and national partners, including space agencies.

While coordination occurs, the two coastal programs have taken somewhat different approaches. The coastal module of GOOS is developing an extensive system linking regional and national programs and has focused on establishing a core observation system and promoting the activities of the individual, generally national, programs (UNESCO, 2003b). The coastal module of GTOS (C-GTOS) has focused on a more limited group of specific issues than GOOS and has separated its implementation into two phases (FAO, 2005). The first phase addresses a set of defined priority proposals. These are designed to provide tests of concept for the observing system and initial important products that can be built upon for the second phase. The second phase is the mature system that complements the coastal module of GOOS, and other coastal observing initiatives, and develops the four phenomena of interest that are the long-term focus issues of C-GTOS identified in the *C-GTOS Strategic Design and Phase 1 Implementation Plan* (FAO, 2005):

- Human dimensions, land use, land cover and critical habitat alteration.
- Sediment loss and delivery.
- Water cycle and water quality.
- Effects of sea level change, storms and flooding.

One of the major challenges in establishing and maintaining a global network for observations of global change is ensuring adequate spatial and temporal extent of the observations. In this article, we describe initial steps toward the development of a global coastal observing system by reviewing (1) the general concerns of observing systems and more specifically coastal in situ observations, (2) the definition and delineation of the coast by potential users, and (3) the application of a sentinel system framework as a means to identify and establishing a network to meet observing system goals. This work stems from one of the phase 1 priority products of the C-GTOS plan "Management of conservation and cultural sites in the coastal zone" (FAO, 2005).

Need for in situ networks and issues in monitoring

Worldwide coverage of observations is key to global observing systems, but for many variables, if not most, such coverage is beyond our current capabilities. Global coverage may exist for satellite and other remote imagery, but the number of variables measured is limited. Not all countries are capable of making sustained, operational, and accessible observations with political, social and economic factors sometimes preventing conducive conditions for these capabilities. The limitations of in situ measurements are particularly apparent; as not all variables can be remotely sensed, and those that can still require some form of in situ validation. Total in situ coverage is lacking no matter what political, social and economic conditions. Even in developed countries the institutional mechanisms for coordinated reporting on in situ monitoring data are far less advanced than those of remote sensing efforts. All things cannot be measured at all places, therefore interpolation is always needed, but the frequency and consistency of measurements is often a limiting factor. Finally, uncertainties of the observations, the derived information, and resultant models are difficult to assess.

International observing systems initiatives, such as GTOS, GOOS, GCOS, IGOS, and the Global Earth Observation System of Systems (GEOSS), provide means to coordinate observing system efforts among countries, assist in implementation and training, and identify gaps for prioritized development and funding. GEOSS is an intergovernmental network that plans to improve observing system significantly over the next 10 years, supported by the efforts the four proceeding initiatives. The observing systems themselves cannot directly make the necessary measurements. Instead, the objectives of these programs will only be achieved through countries contributing observations from their existing or proposed program activities and thus funding. In turn, the observing systems and international donor community must facilitate new measurements where these are lacking, and play an essential role in the coordination, analysis and communication of regional and global information.

The sentinel system concept and its application to global and regional observations

How then can observations be made efficaciously? A long-term goal of observing systems is to build capacity for all nations to make and report observations, but this requires significant commitment and time. The ability to reach the necessary capacity for all nations will be measured in decades at best, so more near-term strategies must be taken. Two such strategies, which are not mutually exclusive, are to (1) develop models that interpolate and extrapolate to areas with little to no data, and (2) observe and build upon "sentinel ecosystems", forming an early warning and core system for broader regional and global change. Sentinel Ecosystems refer to a limited number of well understood systems that have substantial datasets and are observed in a sustained fashion. Jassby (1998) previously defined sentinel ecosystems in another context and considered the implications of the concept with three aquatic ecosystems. The framework for his site selection was based on the following criteria:

- "... relatively small number of locations are chosen for intensive study, each location being selected to represent a certain, preferably large, class of ecosystems."
- "some subset ... must encounter the stressor,"
- "some [sites] ... must have the critical features that cause responsiveness to that stressor,"
- "background variability ... must not disguise the response to the stressor of interest."

With some modification, Jassby's sentinel system framework has application for observing systems of global or regional change, although this was not its original purpose. Furthermore, the sentinel system concept has many advantages in the short- and long-term development of observing systems; it is complementary to the site selection methodologies and policy frameworks of existing programs that the coastal observing system will draw upon, and addresses issues of scale and hierarchy of processes to be observed.

A sentinel ecosystem framework for global observing systems may have several features added to those identified by Jassby, advantageous to the assessment of global and regional change:

- These ecosystems may be both natural and human influenced.
- Where possible, they should meet the criteria outlined in Jassby's approach (both representative of classes of ecosystem and responsiveness to stressors). Although, different approaches will be required regionally, with greater flexibility in the application of criteria for sites in countries with limited resources.
- These systems will likely be important enough for society to support long-term study. Systems important to human society often have historic records and commitment to ongoing study—both for research and monitoring.
- Their value may be for conservation, natural heritage, cultural and socio-economic purposes, providing a diversity of potential assessment capabilities.
- Lastly, these sites are likely to be already part of at least one international network of sites, providing mechanisms for information access.

There are several advantages to this expanded strategy that contribute to the development and management of observing systems, as well as the individual sites themselves. Firstly, the sentinel ecosystems are directly useful for long-term observations of the particular location and representative for assessment of a broader group of ecosystems. Secondly, the work builds capacity for providing sustained and quality-assured information for the development, validation, and evaluation of large scale modeling and comparative change studies (Rastetter et al., 2003). Once the capacity is established, it can be extended; models can be transferred with greater confidence to other locations and adapted for use in regional and global studies (Seitzinger & Kroeze, 1998; Alexander et al., 2002; Seitzinger et al., 2002; Rastetter et al., 2003). Thirdly, this also builds sound region-specific understanding of coastal systems to support regionalized modeling efforts, advocated in advanced global coastal change programs (Church, 2001). Finally, this approach bolsters existing networks of monitoring sites and their activities; it links networks with vested interests in sustained monitoring, and supports current global and regional program activities for monitoring and conservation of heritage areas. These are all necessary steps for global change assessment.

We consider sentinel sites as ecosystems and use the terms somewhat interchangeably. This recognizes a broad definition of ecosystems (Christian, 2003a) that includes the hierarchical nature of observing systems and their components. Sentinel ecosystem observations provide various links to global assessment. Hallmarks of ecosystem studies are elemental cycling and trophic dynamics, both important to global change. In observing ecosystems, often other hierarchical levels of interest are considered (Allen & Hoekstra, 1992; Christian, 2003a). Community structure and population dynamics of keystone and "flagship" species may be assessed in the broader, ecosystem-level context. In turn, a context for ecosystem dynamics and structure is the landscape, perhaps watershed, condition. Global assessment is at least partially derived from some metric (e.g., sum, mean, median) of the conditions of the populations of individual locations around the world. The sentinel ecosystems represent important, first-order sites to sample from these populations.

Definition and delineation of coastal areas for a coastal observing system

Defining the coastal system is an early and obvious task for such efforts as the development of a GTOS coastal observing module. C-GTOS (FAO, 2005) reviewed the definitions and approaches adopted by both potential users and collaborators dealing with coastal issues, grouping these as either multilateral environmental agreements and initiatives, international organizations, and global/regional/national assessments (FAO, 2005). A range of definitions of coastal areas and ecosystems were obtained, showing increasing complexity over time (Table 1). Earlier definitions focused on geographic boundaries and management units (e.g., Exclusive Economic Zone [EEZ] and various Integrated Coastal Area Management [ICAM] guidelines). More recent coastal management initiatives have had a greater focus on ecosystem functionality and include interaction with human use dynamics. The ecosystem approach as endorsed by the Convention on Biological Diversity (CBD), the Millennium Ecosystem Assessment (MA), and other current global assessments and initiatives are consistent with this (Millennium Assessment, 2003). These approaches are not mutually exclusive. Accordingly, C-GTOS has not adopted a definition of the coastal zone in a single way, but will use an adaptive approach specific to the user needs of products to be developed.

In the application of the sentinel site framework for developing a coastal observing system, multiple definitions of the coast are desirable, encompassing as many existing program "coastal" sites, ecosystem types, management zones, and monitoring activities as possible (Table 1). The four focus issues of C-GTOS mentioned earlier will be evaluated within different geographic delineations. For example, "Human dimensions, land use, land cover and critical habitat alteration" might be considered over a larger area than "Effects of sea level change, storms and flooding". Furthermore, the selection

Table 1 Definitions of coastal areas and ecosystems used by international initiatives with coastal mandates (direct quotes are shown in italics)

International initiatives with coastal mandates	Definitions of coastal areas or associated ecosystems and habitats
Multilateral environmental agreements and initiatives	
The United Nations Millennium Assessment (MA) is an international work program designed to meet scientific information needs concerning the consequences of ecosystem change and available options for response. Documentation: Millennium Assessment (2003); http://www.millenniumassessment.org/.	The Millennium Assessment reports on ecosystems and ecosystems services within six reporting categories including the coastal zone. Each category is defined by (i) a central concept and (ii) boundary limits for mapping. *Central Concept: interface between ocean and land, extending seawards to about the middle of the continental shelf and inland to include all areas strongly influenced by the proximity to the ocean. Boundary Limits for Mapping: area between 50 m below mean sea level and 50 m above the high tide level or extending landward to a distance 100 km from shore. Includes coral reefs, intertidal zones, estuaries, coastal aquaculture and sea grass communities.* MA reporting categories are not mutually exclusive: *a wetland ecosystem in a coastal region may be examined both in the MA analysis of coastal systems as well as in its analysis of inland water systems.* Differentiation is made between the coastal zone and other adjacent reporting categories based on the definition of boundary limits for mapping. For example, the coastal zone has a shared boundary with bordering marine systems (*>50 m depth*). Permanent inland waters of inland water systems are also separated spatially from respective coastal systems (*permanent water bodies inland from the coastal zone*).
The Ramsar Convention on Wetlands held in Ramsar, Iran, in 1971, covers all aspects of wetland conservation, recognizing wetlands' importance for biodiversity conservation and the well-being of human communities. Documentation: Ramsar Convention on Wetlands (1971) and associated key documents (Articles 1.2 and 2.1); http://www.ramsar.org/.	The Ramsar definition of wetlands accounts for a wide variety of coastal habitats. The Ramsar Classification System for Wetland Type lists the following types of coastal wetlands: permanent shallow marine waters; marine subtidal aquatic beds; coral reefs; rocky marine shores; sand, shingle or pebble shores; estuarine waters; intertidal mud, sand or salt flats; intertidal marshes; intertidal forested wetlands; coastal brackish/saline lagoons; coastal freshwater lagoons, and karst and other subterranean hydrological systems. The Convention on Wetlands describes wetlands as: *areas of marsh, fen, peatland or water, whether natural or artificial, permanent or temporary, with water that is static or flowing, fresh, brackish or salt, including areas of marine water the depth of which at low tide does not exceed six meters.* [Wetlands] *may incorporate adjacent riparian and coastal zones, islands or bodies of marine water deeper than six meters at low tide lying within the wetland.*
Agenda 21 was adopted at the United Nations Conference on Environment and Development (UNCED) in Rio de Janeiro, Brazil (1992). It is a key integrated coastal area management document and led the way for subsequent coastal area agreements and legal instruments. Documentation: UNCED (1992); http://www.un.org/esa/sustdev/.	Chapter 17 includes seven major program areas that relate to coastal areas and management with some definition of the coast. The first *integrated management and sustainable development of coastal areas, includes the area covered by Exclusive Economic Zones.*

Table 1 continued

International initiatives with coastal mandates	Definitions of coastal areas or associated ecosystems and habitats
International organizations	
The United Nations Environment Programme (UNEP) is developing a the Assessment of the Coastal and Marine Environment (CME) as a contribution to the planned Global Marine Assessment (GMA). This expands upon existing assessment initiatives coastal and marine ecosystems. Multiple other coastal-related initiatives have been conducted, such as the program on Integrated Coastal Area and River Basin Management (ICARM) relevant to the terrestrial coast. Documentation: UNEP (2004); UNEP/MAP/PAP (1999); http://www.unep-wcmc.org/marine/.	An exact definition and spatial extent is not specified for coastal habitats that are part of the CME assessment. Instead an adaptable approach is proposed to determine the scope, based on existing assessment methodologies: *the geographical structure of the assessment has to be flexible and based on natural, political and institutional realities. Existing geographical and programmatic structure ...should be used where appropriate.* The large variety of habitats in coastal waters is noted, including *coastal wetlands, estuaries and deltas, mangrove, coastal reef and seagrass beds.* ICARM guidelines identify the area of concern as encompassing *the catchment, the coastal zone and the near-shore coastal waters...Four interacting zones are taken into consideration: coastal waters, the coastal strip, estuary, and the coastal plain.*
The United Nations Education Scientific and Cultural Organization (UNESCO) has numerous coastal initiatives relating to coastal assessments taking place primarily through the Intergovernmental Oceanographic Commission (IOC), which (as with many UN coastal initiatives) collaborates routinely with Small Island Developing States (SIDS). Integrated Coastal Area Management (ICAM) is one such program, which is currently developing indicators for assessment of the coastal area. Documentation: UNESCO (2003a); http://www.ioc.unesco.org/.	A guide published on the use of indicators for ICAM states that catchment management deals with land usages in the coastal stream and river runoff areas for lagoons, bays and estuaries.
The Food and Agriculture Organization (FAO) of the United Nations has multiple initiatives addressing coastal areas, their management and the production of relevant guidelines, such as the Code of Conduct for Responsible Fisheries. Documentation: Scialabba (1998); FAO (1995); http://www.fao.org/.	The FAO ICAM guidelines state: *an ICM program embraces all of the coastal and upland areas, the uses of which can affect coastal waters and the resources therein, and extends seaward to include that part of the coastal ocean that can affect the land of the coastal zone. The ICM program may also include the entire ocean area under national jurisdiction (Exclusive Economic Zone), over which national governments have stewardship responsibilities under both the Law of the Sea Convention and UNCED.*
Global/national assessment initiatives	
The Coastal Ocean Observations Module of the Global Ocean Observing System (C-GOOS) has been developed with the goal of monitoring, assessing, and predicting the effects of natural variations and human activities on the marine environment and ecosystems of the coastal ocean. Documentation: UNESCO (2003b); http://www.ioc.unesco.org/goos/coop.htm.	Coastal, as defined for use in the Coastal Module of GOOS, refers to *regional mosaics of habitats including intertidal habitats (mangroves, marshes, mud flats, rocky shores, sandy beaches), semi-enclosed bodies of water (estuaries, sounds, bays, fjords, gulfs, seas), benthic habitats (coral reefs, sea grass beds, kelp forests, hard and soft bottoms) and the open waters of the coastal ocean to the seaward limits of the Exclusive Economic Zone (EEZ)*, i.e. from the head of the tidal waters to the outer limits of the EEZ. The definition of coastal zone is adopted from Nicholls and Small (2002): *the land margin within 100 km of the coastline or less than 100 m above mean low tide, which ever comes first.*

Table 1 continued

International initiatives with coastal mandates	Definitions of coastal areas or associated ecosystems and habitats
The International Geosphere-Biosphere Project's (IGBP) mission is to deliver scientific knowledge to help human societies develop in harmony with earth's environment. The mandate of Land-Ocean Interactions in the Coastal Zone (LOICZ), as a core project of IGBP, is to address global change in coastal systems and to inform earth system sciences on the relevance of global change in coastal systems. Documentation: IGBP Secretariat (2004); http://www.loicz.org/.	LOICZ includes in its statement of major goals the following reference to the costal zone and scales of activity: *to provide a framework...and to act as a means to focus on key issues concerning human activity and resource use in the coastal zone by applying the full water-continuum scale including the river catchments and the EEZ as spatial scales of major human interventions.*
The Global International Waters Assessment (GIWA) assesses international waters and associated basins, providing needed information for related Global Environment Facility (GEF) activities. A GEF objective for this focus area is to serve primarily as a catalyst to the development of a more comprehensive, ecosystem-based approach to managing international waters and their drainage basins. Documentation: Pernetta and Mee (1998); UNEP (1999); http://www.giwa.net/; http://www.gefweb.org/.	International waters and their drainage basins, which include coastal areas, are one of four priority areas identified by GEF and assessed by GIWA. These combined areas include different coastal habitats *comprising marine, coastal and freshwater areas, and surface waters as well as groundwaters.* The determining factor for this geographic delineation was the *integrity of each unit in terms of encompassing the major causes and effects of environmental problems associated with each transboundary water area, whether river basin, groundwater, lake or sea. In many cases, a drainage area and associated marine basin (often a large marine ecosystem, LME) were the most appropriate units.*
The World Conservation Union (IUCN), National Oceanic and Atmospheric Administration (NOAA), and other organizations that assist developing countries in implementing ecosystem-based strategies use Large Marine Ecosystem (LME) as the principal assessment and management units for coastal ocean resources. Documentation: Sherman and Duda (1999); http://www.iucn.org/; http://www.noaa.gov/.	LMEs include multiple coastal habitats as they *are regions of ocean space encompassing coastal areas from river basins and estuaries to the seaward boundary of continental shelves and the outer margins of the major current systems. They are relatively large regions...characterized by distinct bathymetry, hydrography, productivity, and tropically dependent populations.*

of sentinel systems ultimately requires investigation of ecosystem functioning and susceptibility to stressors, both of which require further analysis that cannot be derived solely by analysis using geographic delineations. We considered a range of definitions, but began with the definition of coastal areas as less than 100 km inland and less than 100 m in elevation (used by Small & Nicholls, [2003] for coastal population mapping), and with a seaward boundary coinciding with the seaward edge of the EEZ. This delineation encompasses the majority of definitions of coastal areas described in Table 1 and serves as a baseline. Moreover, it is used by the coastal program of GOOS (UNESCO, 2003b) and other planned products for C-GTOS (FAO, 2005). Thus, elevation and distance from coast are two geographic parameters important for defining the coastal ecosystems identified in Table 1, and represent an initial step towards determining key geomorphologic characteristics of potential sites.

Application of the sentinel system framework to identify networks and sites

A coastal observing system must build upon existing initiatives, many of which share sites. They often have complementary, but different, monitoring activities and reasons for site selection. Some networks and programs to draw from include the International Long-term Ecological Research (ILTER), Man and the Biosphere, The Nature Conservancy, the Ramsar Convention on Wetlands and the World Heritage Convention. Many of these initiatives have goals in common with the use of the sentinel system framework for coastal observing system site selection, both in the process of selection of sites and ongoing support of networks. For example, the US Long-Term Ecological Research Program (LTER) has a strategy to develop long-term and large spatial scale research (Kaiser, 2001; Hobbie et al., 2003). Similar frameworks have also been developed for the ILTER, the International counterpart of LTER, to identify appropriate research questions for regional ecosystem investigations (Christian & Gosz, 2001; Rivera-Monroy et al., 2004). The framework developed by Rivera-Monroy et al. (2004) uses the environmental signature hypothesis to develop a series of research questions for the reef-seagrass-wetland seascape of the Caribbean region. Like the sentinel ecosystem framework, geophysical processes and biogeochemical properties of a small number of different sites will be used to investigate how stressors limit the function of representative ecosystems. TEAM (Tropical Ecology, Assessment and Monitoring) is another initiative compatible with the sentinel system framework. TEAM, led by Conservation International, will support a network of field stations in tropical areas to monitor long-term trends in biodiversity, providing an early warning system on biodiversity change. The selection of TEAM sites will build on existing Conservation International activities, identifying tropical biodiversity hotspots and important wilderness area requiring monitoring (For more information on TEAM, see http://www.teaminitiative.org).

Difficulties are confronted when selecting sentinel sites that were initiated through other programs for a number of reasons. These sites may not have been selected as representative of a particular ecosystems or geographic location. Often sites were selected for pragmatic reasons; sites may have been of local importance, unique in some way, or just more (or less) accessible. Jassby (1998), therefore, states that the problem of selecting sentinel systems is not so much where to put sites, but rather how to analyze better existing sites to serve the needs of the resource assessment. In the context of C-GTOS, this requires the identification of sites that best represent the underlying mechanisms of response to global, large-scale and/or ubiquitous environmental drivers (FAO, 2005). Access and continuity of long-term observations are also key factors to identifying sites that can sustain the types of measurements required. Nations and international bodies with vested interests may then be asked to contribute to the observing system by focusing attention on a limited number of these key sites that can realistically be supported for long-term and coordinated observations. Ultimately the success of the observing system will reflect countries' commitment to supporting these initiatives.

Methods

We identified potential coastal sentinel sites from existing international programs. Sites were also examined to investigate the applicability of various coastal area delineations for coastal site selection. A number of steps were required to achieve this, described in the following sections.

Definition and delineation of coasts

Geographic Information System (GIS) data files were created for the various global coastal areas used throughout the study. These were generated using two data sources: the UN Cartographic Section (UNCS) country boundaries map of the world (1:1,000,000 scale), and the United States Geological Survey (USGS) Digital Elevation Model (GTOPO30) which has a resolution of 30 arc seconds. All coastal area maps were developed by the Environment and Natural Resources Service (SDRN) of the Food and Agriculture Organization (FAO) for the coastal population product of C-GTOS (FAO, 2005). These were based on datasets originally prepared for the Global Poverty Mapping Project of SDRN. A baseline coastal area delineation was developed including all land areas globally less than 100 km inland and 100 m in elevation. Four other global delineations were created, each with an increasing area of land, or buffer, from the ocean-land boundary (5 km, 10 km, 50 km, and 100 km).

Selection of coastal program sites using the sentinel system framework

Site locations and associated data were collected for programs that have monitoring or protected area sites. Data files of site locations were obtained for each organization where available, including publicly available GIS files, online databases, and publications and reports. All available geo-referenced site data were incorporated into a GIS, and the number of coastal sites was then determined for each program or initiatives using the coastal delineations created. Where geo-referenced data were not available, the number of coastal sites was calculated from information published in other sources, based on the program's classification of coastal sites or environments.

All coastal area delineations were used in further analysis of the suitability of various coastal definitions for selecting coastal sentinel ecosystem sites. To achieve this, the number of sites for some of the programs was recalculated using all delineations, and the resulting number of sites falling within these areas was compared. In addition, site data registered in TEMS (Terrestrial Ecosystem Monitoring Sites) was further analyzed. TEMS is the online database of GTOS and contains site metadata information (Tschirley et al., 2003), including information on which sites monitor variables applicable to the four "phenomena of interest", or focus issues, identified for C-GTOS (FAO, 2005):

- Human dimension, land cover/land use and critical habitat alteration (37 variables).
- Sediment loss and delivery (22 variables).
- Water cycle and water quality (27 variables).
- Effects of sea level, storms and flooding (13 variables).

We analyzed the TEMS data to determine whether one or more variables relevant to C-GTOS are monitored at each of the sites found within each of the coastal delineations (FAO, 2005). If a site did contain at least one variable of importance to a particular C-GTOS focus issue, the site was considered in the total count of potential sites relevant to that issue. Therefore, the number of potential sites identified within each coastal delineation could be compared for each of the focus issues.

Results

Definition and delineation of coasts

A GIS project was created that contained all of the geo-referenced sites examined and all coastal area delineations developed. Figure 1 represents an example image from the complete GIS showing the location of Ramsar Convention on Wetlands sites and coastal areas delineations for areas less than 100 m in elevation, and 50 km and 100 km from the coast. Although difficult to identify coastal areas less than 100 m in elevation at the scale depicted in the Fig. 1, this does highlight the expanse of coastline where coastal lowlands (areas of low elevation) do not extend very far inland towards the 100 km boundary. The exceptions to this are delta areas, such as the Nile Delta shown in Fig. 1.

Selection of coastal program sites using the sentinel system framework

Each of the 10 programs and initiatives investigated have some proportion of sites or records of monitoring activities in coastal areas (Table 2). Three of the 10 programs have sites recognized as areas of conservation or cultural significance with varying requirements for ongoing monitoring, they are: Ramsar Convention on Wetlands (Ramsar), United Nations Educational, Scientific and Cultural Organization's Man and the Biosphere (UNESCO-MAB), and UNESCO's World Heritage Convention (UNESCO-WH). The remaining seven initiatives have coastal sites or activities that primarily monitor environmental change (Table 2). With the exception of the IUCN (World Conservation Union) Red List, which reports species distribution, all initiatives have fixed site locations varying from a few meters to 1000's of hectares in size, and are therefore suitable for consideration as sources of sentinel sites.

GIS data files with the location of sites for Ramsar, UNESCO-MAB, and UNESCO-WH

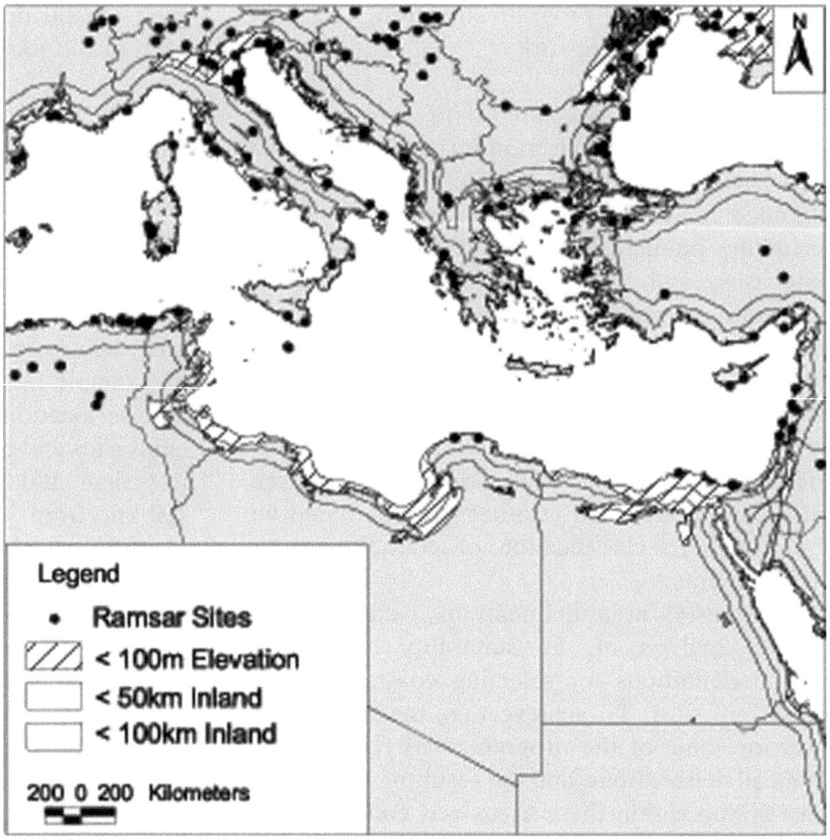

Fig. 1 A GIS map showing project created showing various coastal area delineations and the location of sites of the Ramsar Convention on Wetlands in countries bordering the Mediterranean

were provided by the UN Environment Programme's World Conservation Monitoring Centre (UNEP–WCMC), available as part of the World Database of Protected Areas (WDPA, 2005). GIS data files or information used for the other seven initiatives investigated were obtained directly from the programs and all sources are indicated in Table 2. The locations of sites of the three initiatives for areas of conservation and cultural importance, as well as all TEMS sites, were further analyzed to compare the results of using various coastal area delineations to identify coastal sites (Table 3). The baseline coastal area delineation of within 100 km and less than 100 m in elevation includes between 22 and 29% of the total number of sites for each program, except for Ramsar in which 56% of its sites are coastal. Using a somewhat broader definition, each program has slightly less than half of their sites within 100 km of the coast, with the exception of Ramsar, which has 66% of sites within the 100 km coastal zone. In fact, proportionally Ramsar has more sites proximal to the coast, with at least two times the proportion of sites within 5 km of the coast when compared to other programs. The majority of these Ramsar sites are in low lying areas which can be seen in the small difference of 137 sites (representing less than 10% difference) identified with the baseline delineation (combined distance and elevation) compared to the 100 km distance delineation alone (Table 3). The coastal proximity and low elevation of Ramsar sites can be inferred from Fig. 1, but is more apparent in the analysis shown in Table 3. The trends for the two UNESCO programs (MAB and WH) are very similar, whereas TEMS generally has the least proportion of coastal sites by most delineations (Table 3).

Further analysis of TEMS site data shows that at least 1 variable is measured at each site from one or more of those listed as relevant to the focus issues of C-GTOS (FAO, 2005). Therefore, each of the 2,168 sites examined has relevance to C-GTOS, but not all sites contain variables

Table 2 Existing global initiatives that have observations or sites within coastal areas, as defined by the baseline coastal area delineation

Organization, website, and source of data used	Number of sites or records of monitoring activities, requirements, and frequency	Coastal sites (% of total sites)
Monitoring sites or activities		
GTOS - Terrestrial Ecosystem Monitoring Sites (TEMS) (http://www.fao.org/gtos/tems/; Christian, 2003b)	2168 sites were examined from the total of 2667[a] registered TEMS sites. All sites are must include monitoring at least every 5 years.	482 (22%)
International Organization of Biological Field Stations (http://www.obfs.org)	290+ sites. No requirement for sites to carry out regular monitoring.	[b]
IUCN Red List (http://www.redlist.org/)	In 2004 Red list 4998 species were identified as critically endangered, endangered, or vulnerable. The Red list is updated annually.	154 (3%)[c]
International Long Term Ecological Research (ILTER) (http://www.ilternet.edu/; Hobbie et al., 2003)	The 251 sites registered in TEMS were examined. ILTER sites exist in 25+ countries.	249 (99%)
Land-Ocean Interactions in the Coastal Zone (LOICZ) (http://www.loicz.org/; Smith et al., 2003)	137 individual sites were included from the dataset examined. There is no requirement for regular monitoring.	135 (99%)
ReefBase (http://www.reefbase.org/)	2608 monitoring sites, all of which are reef sites.	2608 (100%)
World Register of Field Sites (http://www.rgs.org/)	377 sites. No requirement for on going monitoring. The register lists field sites and indicates availability of facilities.	17 (5%)
Sites of conservation and cultural value		
Ramsar (http://www.ramsar.org/)	1384[d] sites. Many sites contain ongoing monitoring activities, but no formal requirement.	770 (56%)
UNESCO—Man and the Biosphere Programme (http://www.unesco.org/mab/)	448[d] sites. Many sites contain ongoing monitoring activities, but no formal requirement.	106 (24%)
UNESCO—World Heritage Convention (http://whc.unesco.org/)	801[d] sites. All sites have some reporting requirements. Many sites contain ongoing monitoring activities, but no formal requirement.	231 (29%)

[a] All sites were excluded that did not have the necessary coordinate or network information for further analysis

[b] Information on site locations was not available to determine the percentage of coastal sites

[c] Information on site locations was not available for GIS analysis. The number of coastal sites was calculated using the available database criteria classifying coastal habitats where the data was collected. Data identified as coastal or sea (excluding open oceans) was used. This figure is more representative of the number of coastal endangered species than sampling effort in coastal areas

[d] Number of sites based on analysis of spatial data publicly available for 2005. The number of sites at the time of submission was: Ramsar (1524), Man and the Biosphere (482), Word Heritage Convention (812). Current official numbers are published in the listed websites

important to all four focus issues (Table 4). The variables for the focus issue "Human dimensions and critical habitat change" are measured at the most sites across all delineations. "Sediment loss and delivery" is the next most observed, followed by "Water cycle and water quality."

Only sites within 5 km of the coast were examined for investigation of the focus issue "Effects of sea level, storms and flooding". Even though sites further inland monitor relevant variables, the information required for this particular C-GTOS focus issue is for the major part localized to near coastlines (e.g. wind speed). Therefore sites more than 5 km inland were excluded and the resulting 291 sites for this focus issue deal with only the coastal specific variables.

Discussion

Definition and delineation of coasts

Inclusive, yet flexible, definitions of coastal areas for C-GTOS must address user needs, and aid in

Table 3 Number of global program sites located within various coastal area delineations. Percentage of total number of program sites is indicated in parenthesis (%)

Coastal area delineation	TEMS	Ramsar	UNESCO-MAB	UNESCO-WH
Within 5 km of the coast	310 (14%)	592 (43%)	98 (22%)	169 (21%)
Within 10 km of the coast	394 (18%)	670 (48%)	121 (27%)	210 (26%)
Within 50 km of the coast	698 (32%)	861 (62%)	177 (40%)	333 (42%)
Within 100 km of the coast	951 (44%)	907 (66%)	197 (44%)	388 (48%)
Less than 100 m in elevation	519 (24%)	803 (58%)	113 (25%)	246 (31%)
Within 100 km of the coast and less than 100 m elevation	482 (22%)	770 (56%)	106 (24%)	231 (29%)
Total number of program sites	2168[a] (100%)	1384 (100%)	448 (100%)	801 (100%)

[a] All sites were excluded that did not have the necessary coordinate information or network information for further analysis. TEMS has a total 2667 registered sites

Table 4 Number of TEMS sites that monitor variables of importance to C-GTOS, located within various coastal area delineations. The number of sites monitoring variables relevant to each of the four focal issues of C-GTOS is included for each of the coastal delineation investigated

Coastal area delineation	Total C-GTOS related TEMS sites	Sites addressing C-GTOS focal issues			
		Human dimensions and critical habitat change	Sediment loss and delivery	Water cycle and water quality	Effects of sea level, storms and flooding[a]
Within 5 km of the coast	310	263	198	198	291
Within 10 km of the coast	394	333	250	242	–
Within 50 km of the coast	698	607	492	374	–
Within 100 km of the coast	951	838	682	499	–
Less than 100 m in elevation	519	455	358	261	–
Within 100 km of the coast and less than 100 m elevation	482	420	327	250	–
Total sites	2168[b]	1957	1536	1152	–

[a] All sites further than 5 km from the coast were not included in the analysis of sites monitoring the effects of sea level, storms and flooding

[b] All sites were excluded that did not have the necessary coordinate information or network information for further analysis. TEMS has a total 2667 registered sites

site selection. We identified the coastal observations addressed by various programs, and determined the number and location of sites in their networks. A broad spatial definition (from the outer edge of the EEZ to 100 km inland) with an elevation component (less than 100 m in height) was used as a baseline. It allowed identification of a reasonable number of potential "coastal" sites (Tables 3 and 4). This delineation encapsulated the majority of definitions of coastal areas used by potential users and information providers to an integrated coastal observing system (Table 1), and is the same as used by the coastal module of GOOS (UNESCO, 2003b). It reflects the reasoning that sites close to the coast but high in elevation are less likely to be influenced by coastal/marine processes than sites in lower elevations. However, some aspects of ecosystems in higher elevations may influence coastal processes. For example, run off from these sites may be important to coastal conditions near these sites. The number of coastal sites defined by 100 km alone was always higher than by the dual factor definition. There was a comparatively small change in the number of Ramsar sites, regardless of if elevation was included as a factor or not (Table 3). Ramsar wetland sites are thus often coastal and found in low lying and near sea areas, while the mandate of other programs covers broader habitats that may not be aquatic.

Selection of coastal program sites using the sentinel system framework

The identification of potential sentinel ecosystems for a coastal observing network was initiated in steps, each with specific assumptions and constraints. First, we identified programs and initiatives with mandates to monitor environmental change (Table 1). Relevant and long-term data were found to be accessible at a large number of sites within these programs (as was identified in Table 4, examining TEMS registered site information and C-GTOS focus issues). Second, we then developed delineations representing widely accepted definitions of coastal areas and used this to identify a subset of potential "coastal" sites from these programs for further investigation. This was found to be crucial step in site selection, as many networks had greatly varying objectives not necessarily coinciding with the needs of C-GTOS; in fact, most initiatives examined had the majority of their sites outside of coastal areas (Table 2). Using this methodology, we focused on 10 specific networks. This approach also proved useful in the interrogation of registered site data in the TEMS database to identify potential coastal sites from other programs and networks (Table 4). Further investigation of registered networks in TEMS will be useful in the selection of sentinel sites, particularly for identification of sites from essential networks such as ILTER. Such networks use TEMS for publication of site, contact, and variable information (for more information see the ILTER website http://www.ilternet.edu/ or TEMS http://www.fao.org/gtos/tems/).

The next steps of selecting sentinel sites will involve closer inspection of the individual sites, direct identification of sentinels among the potential coastal sites, and formal incorporation into a sentinel network. This process will be informed and driven by the joint efforts of the various programs and initiatives, a number of which have been contacted to begin this process, including LOICZ, UNESCO-WH, UNESCO-MAB, and UNEP-WCMC.

Joint efforts between programs using sentinel systems

All networks of sites examined are potential partners in GTOS. The objectives of site monitoring and the networks to which they belong vary widely, but have three broad classes of predominant function, useful in considering collaborative efforts:

- programs with sites of recognized conservation and cultural value;
- programs with sites for monitoring change;
- programs carrying out global and regional assessments of change.

Some collaborative efforts have been initiated within and, to a lesser extent, between these different types of initiatives for some time. Further efforts focused specifically on coastal issues are needed. The following briefly details some of the current status and opportunities regarding this, and identifies how the sentinel system approach may support these actions.

Efforts have been made, through a number of different initiatives to strengthen links between the three programs of conservation and cultural sites examined (Ramsar, UNESCO-MAB, and UNESCO-WH). The most substantial of these initiatives is the mapping of all program site locations through a single geo-referenced World Protected Areas Database, led by UNEP-WCMC. Some efforts have also been made to identify sites in common between the three programs. Eighteen sites where found to be in common between all three programs, 101 sites are jointly MAB and Ramsar, and 75 are in both UNESCO-MAB and UNESCO-WH (for more information see the lists of sites identified on the MAB website: http://www.unesco.org/mab/wnbr.htm). Of these sites, many fall within those identified as coastal in this study and may serve as a good starting point for identification of specific sentinel ecosystems.

Regional and global initiatives that coordinate activities between networks will be instrumental in identifying and synthesizing site data needed for global and regional assessment and observing system programs. This includes initiatives such as GTOS-TEMS, The International Register of

Field Stations (Table 2), and the World Register of Field Sites. Even with such efforts, further collaboration directly between international programs with vested interests in long-term monitoring is necessary for the programs themselves, and for the development of a sustained integrated program of long-term monitoring of coastal change (Kaiser, 2001; Hobbie et al., 2003; Smith et al., 2003).

Many global initiatives, such as the IUCN Red list, use a combination of data from monitoring activities, modeling, and expert opinion to derive global and regional assessment products (Table 2). Such initiatives are potential users of the proposed global network of coastal observation sites and resulting GTOS products. It is not necessarily the case that data used in assessments are derived repeatedly from the same areas or sites. This being said, ongoing global and regional assessments are also obvious partners in developing a network of sentinel sites for a number of reasons:

- Many assessment initiatives do have subsets of data and ongoing monitoring locations that may be relevant to selected coastal sentinel ecosystems;
- Monitoring activities at sentinel ecosystem sites are designed to contribute to regional and global assessments, and collaborative efforts between programs will aid in prioritize these efforts, and save resources;
- There is a large overlap in the network of experts needed for each, at both country and regional levels.

The global assessment initiatives that should be considered include the following (for more information see Table 1 and listed web references below):

- the Millennium Assessment and subsequent regional assessments;
- IUCN Red List, and its partner organizations such as the World Fish Center and FAO that contribute coastal data through programs such as FishBase and the FAO Species Identification and Data Program;
- UNEP activities such as the Global Marine Assessment, Global Environmental Outlook and Global International Waters Assessment;
- coastal and marine assessments of non profit organizations such as World Resources Institute (http://marine.wri.org/projects.cfm), The Nature Conservancy (http://nature.org/) and Conservation International (http://www.conservation.org).

UNEP-WCMC has a marine program active since 2000, and has produced global and regional coastal mapping and assessments products including atlases of mangroves, seagrasses, and coral reefs. At the time of publication, WCMC were in the final stages of completing a Caribbean regional initiative, collecting fisheries data in relation to marine protected areas (E. McManus, UNEP-WCMC, pers. comms.). Activities such as these are key to the development of a coastal sentinel ecosystem network; they bring together assessment activities, monitoring data, and sites of conservation and cultural importance in the coastal zone.

Sentinel ecosystems as a strategy for global and regional observing systems

The Sentinel ecosystem strategy benefits many of the short- and long-term goals of observation programs and national custodians responsible for supporting the day to day activities and on-going management of sites. This approach has several advantages. First and most obvious is that the sentinel ecosystems are directly useful for long-term observations of the particular location and representative for assessment of a broader group of ecosystems. For example, a variety of wetland types have been identified and protected through the Ramsar Convention on Wetlands, and the monitoring of many is integrated into the site management plan. The monitoring efforts vary in intensity, and sentinel ecosystems might be chosen from the most active sites. Monitoring can be promoted in others by leveraging the needs of both Ramsar and GTOS.

Second, implementation of the strategy also builds capacity for providing sustained and quality-assured information for the development, validation, and evaluation of large-scale modeling and comparative change studies (Rastetter et al., 2003). Simple models of water and salt balance, biochemical stoichiometry, and nutrient exchange have been constructed through Land-Ocean Interactions in the Coastal Zone (LOICZ) at over 200 sites (Smith et al., 2003). A limited number of these sites may also be characterized by much more sophisticated hydrodynamic, ecological process and/or water quality models. These sites may act to validate and improve the simpler models. Modeling efforts can then be transferred to support regionalized modeling efforts advocated in advanced global coastal change programs (Church, 2001).

Third, the strategy links networks with interests in sustained monitoring and conservation of heritage areas. Our approach is to choose potential sentinel ecosystems from these existing networks.

Thus, the application of a sentinel ecosystem framework will be of benefit throughout the development of an in situ coastal observation network; from the process of site selection, to the management of individual sites and coordination of the operable network. Once site selection is complete and stakeholder commitment reached, many of the activities begin that ensure a sustained operational network, including: gap analysis and establishment of required sites; securing of long-term financial and logistical support; development of standards and harmonization of data, inter-calibration exercises, and guidelines for reporting and management. C-GTOS is now taking the next steps in sentinel ecosystem site selection, identifying specific sites, contacting appropriate authorities, and beginning the process of organizing activities among stakeholders supporting sites of long-term coastal monitoring.

Concluding remarks

We have assessed user-based definitions of the coast and used them to develop delineations for the identification of potential sites for coastal observing systems. Sites within international networks represent potential sentinel ecosystem sites. The sentinel ecosystems framework is a strategy for the development of a sustainable network of sites for global observing system needs; the approach will support the development of early warning systems for broad regional and global change, and selected individual sites can form a backbone for comprehensive observations and study. Sentinel ecosystems have relevance to the development and coordination of coastal observation efforts, and in particular to C-GTOS—addressing the observing system's goals of detecting, assessing, and predicting change in coastal terrestrial, wetland and freshwater ecosystems (FAO, 2005). This strategy not only addresses the goals of observation programs, but perhaps most importantly, the needs of countries and national custodians inevitably responsible for supporting the day to day activities and on-going management of sites.

Acknowledgements This manuscript builds on the one submitted to the proceedings of the CORILA conference on coastal wetlands in 2004. The former manuscript was co-authored by Stephen deMora, and we thank him for his efforts in the early development of these ideas. We thank the organizers of the CORILA conference for providing such a wonderful venue, and for R. Christian's travel support. Further, we thank the C-GTOS expert panel members and GTOS staff for their contributions to the ideas developed here. Particular thanks go to Dan Baird, Bob Bowen, Dave Clark, Howard Diamond, Paul DiGiacomo, Jorge Jimenez, John Kineman, John Latham, Liana McManus, Stephen deMora, Geraud Servin, Pierluigi Viaroli, and Helen Yap. Funding for C-GTOS has come through FAO, GTOS and the U.S. Global Climate Observing System Program of NOAA/NESDIS. Particular thanks go to Ergin Ataman and Mario Bloise of the GIS Group of SDRN (FAO of the UN) for the production of all coastal delineation datasets used within the study. We wish to express our gratitude for the efforts of Aurelien Letourneau providing and analyzing the GTOS TEMS data, Antonio Martucci, Emelie Healy, and Paolo Prosperi for GIS analysis, and Angas Hopkins for his input reviewing the manuscript.

References

Alexander, R. B., P. J. Johnes, E. W. Boyer & R. A. Smith, 2002. A comparison of models for estimating the riverine export of nitrogen from large watersheds. Biogeochemistry 57/58: 295–339.

Allen, T. F. H. & T. W. Hoekstra, 1992. Toward a Unified Ecology. Columbia University Press, New York.

Christian, R. R., 2003a. Concepts of ecosystem, level, and scale. In Bodini A. (ed.) Encyclopedia of Life Support Systems (EOLSS). EOLSS Publishers, Oxford.
Christian, R. R., 2003b. Coastal initiative of the global terrestrial observing system. Ocean and Coastal Management 46: 313–321.
Christian, R. R. & J. Gosz, 2001. Long-term ecological research in concepts and practice. Congresso Nazionale della Società Italiana di Ecologia ATTI 25: 25–31.
Church, J. A., 2001. Climate change: How fast are sea levels rising? Science 294: 802–803.
FAO, 1995. Code of conduct for responsible fisheries. Twenty-eighth Session of the FAO Conference, 31 October 1995. FAO, Rome (available at http://www.fao.org/DOCREP/005/v9878e/v9878e00.htm).
FAO, 2005. Coastal GTOS draft strategic design and phase 1 implementation plan. By Christian, R. R., D. Baird, R. E., Bowen, D. M. Clark, P. M. DiGiacomo, P. M., S. de Mora, J. Jiménez, J. Kineman, S. Mazzilli, G. Servin, L. Talaue-McManus, P. Viaroli & H. Yap. GTOS Report No. 36. FAO, Rome (available at http://www.fao.org/gtos/gtospub/pub36.html).
Hobbie, J., S. Carpenter, N. B. Grimm, J. R. Gosz & T. R. Seastedt, 2003. The US long term ecological research program. Bioscience 53: 21–32.
IGBP Secreteriat, 2004. LOICZ science plan and implementation strategy structure (draft). In Kremer, H. H., M. D. A. Le-Tissier, P. R. Burbridge & C. J. Crossland, (eds) IGBP Secretariat, Stockholm.
Jassby, A. D., 1998. Interannual variability at three inland water sites: implications for sentinel ecosystems. Ecological Applications 8: 277–287.
Kaiser, J., 2001. An experiment for all seasons. Science 293: 624–627.
Millennium Assessment, 2003. Ecosystems and Human Well-Being: A Framework For Assessment. Island Press, Washington DC.
Nicholls, R. J. & C. Small, 2002. Improved estimates of coastal population and exposure to hazards. EOS, Transactions, American Geophysical Union 83: 301.
Pernetta, J. C. & L. D. Mee, 1998. The global international waters assessment. In ACOPS (eds) Towards Enhanced Ocean Security Into the Third Millennium. Advisory Committee on the Protection of the Sea, London, 201–221.
Ramsar, 1971. The Ramsar Convention on Wetlands (available at http://www.ramsar.org/key_conv_e.htm).
Rastetter, E. B., J. D. Aber, D. P. C. Peters, D. S. Ojima & I. C. Burke, 2003. Using mechanistic models to scale ecological processes across space and time. BioScience 53: 68–76.
Rivera-Monroy, V. H., R. R. Twilley, D. Bone, D. L. Childers, C. Coronado-Molina, I. C. Feller, J. Herrera-Silveira, R. Jaffe, E. Mancera, E. Rejmankova, J. E. Salsbury & E. Weil, 2004. A conceptual framework to develop long-term ecological research and management objectives in the wider Caribbean region. BioScience 54: 843–856.
Scialabba, N. (ed.), 1998. Integrated Coastal Area Management and Agriculture and Forestry and Fisheries. FAO guidelines. FAO, Rome.
Seitzinger, S. P. & C. Kroeze, 1998. Global distribution of nitrous oxide production and N inputs in freshwater and coastal marine ecosystems. Global Biogeochemistry Cycles 12: 93–113.
Seitzinger, S. P., R. V. Styles, E. W. Boyer, R. B. Alexander, G. Billen, R. W. Howarth, B. Mayer & N. Breemen, 2002. Nitrogen retention in rivers: Model development and application to watersheds in the northeastern U.S.A. Biogeochemistry 57/58: 199–237.
Sherman, K. & A. M. Duda, 1999. An ecosystem approach to global assessment and management of coastal waters. Marine Ecological Progress Series 190: 271–87.
Small, C. R. J. & A. Nicholls, 2003. Global analysis of human settlement in coastal zones. Journal of Coastal Research 19: 584–599.
Smith, S., D. Swaney, L. Talaue-McManus, J. Bartley, P. Sandhei, C. McLacghlin, V. Dupra, C. Crossland, R. Buddemeier, B. Maxwell & F. Wulff, 2003. Humans, hydrology, and the distribution of inorganic nutrient loading to the ocean. BioScience 53: 235–45.
Tschirley, J., G. Servin & R. Sessa, 2003. The terrestrial ecosystem monitoring sites: a utility assessment. International Journal of Global Environmental Issues 3: 369–381.
UNCED, 1992. Agenda 21. Chapter 17 (available at http://www.un.org/esa/sustdev/documents/agenda21/english/agenda21chapter17.htm).
UNEP, 1999. GIWA project document. Document No. GF/FP/1100–99–01. UNEP, Nairobi.
UNEP, 2004. Towards a UNEP module for the assessment of the coastal and marine environment, draft document (available at http://www.unep.org/DEWA/water/MarineAssessment/GMANBI_meeting/Background%20paper%20-%20UNEP%20Module.doc).
UNEP/MAP/PAP, 1999. Conceptual Framework and Planning Guidelines for Integrated Coastal Area and River Basin Management. Priority Actions Programme, Split (available at http://www.ucc-water.org/Freshco/Docs/ICARM-Guidelines.pdf).
UNESCO, 2003a. A reference guide on the use of indicators for integrated coastal management. ICAM dossier 1, IOC manuals and Guides No. 45. UNESCO, Paris (also available at http://ioc.unesco.org/icam/files/Dossier.pdf).
UNESCO, 2003b. The integrated strategic design plan for the coastal observations module of the Global Oceanographic Observing System. IOC/INF-1183. GOOS Report No. 125. UNESCO, Paris.
WDPA, 2005. World Database on Protected Areas 2005 – Copyright World Conservation Union (IUCN) and UNEP-World Conservation Monitoring Centre (UNEP-WCMC), Cambridge (Available at http://sea.unep-wcmc.org/wdbpa/consortium.cfm).

Hydrobiologia (2007) 577:71–85
DOI 10.1007/s10750-006-0418-3

LAGOONS AND COASTAL WETLANDS

Distribution and production of macrophytes and phytoplankton in the lagoon of Venice: comparison of actual and past situation

Adriano Sfriso · Chiara Facca

Abstract This paper aims at comparing inedited maps of macrophytes and phytoplankton distribution in the lagoon of Venice in 1980 and 2003. The macrophyte distribution is displayed with reference to different biomass intervals which allow the calculation of the occupied surface, standing crop (SC), net (NPP) and gross (GPP) production. In 1980 the total macroalgal SC was ca. 841 ktonnes whereas the annual NPP and GPP were estimated to be ca. 2912 and 18498 ktonnes, respectively. In 2003 macroalgae displayed a marked regression and the SC, NPP and GPP decreased to ca. 89, 471 and 2336 ktonnes. Maps of the seagrass distribution date back to 1990 but their biomass and production have been quantified only in 2003. On the whole, in 2003, on a surface of ca. 56 km^2 the three species living in the lagoon accounted for a SC and a NPP of ca. 209 and 821 ktonnes, whereas the GPP estimated according to literature was about twice as high as the NPP. *Cymodocea nodosa* was the most abundant phanerogam in the lagoon. Its SC was higher than the total of the macroalgae, although the latter were more productive. *Zostera marina* covered the highest surface in the lagoon but its biomass and production were a little lower than that of *C. nodosa*. *Nanozostera noltii*, which was common and widespread in the past, displayed an evident regression trend which was mainly due to the increase of the water turbidity and the disruptions of its habitat. Similarly, phytoplankton, underwent a descending trend, although data quoted in the present paper display its distribution only in the summer period.

Keywords Macroalgae · Seagrasses · Phytoplankton · Standing crop · Primary production · Venice lagoon

Guest editors: P. Viaroli, P. Lasserre & P. Campostrini
Lagoons and Coastal Wetlands in the Global Change Context: Impacts and Management Issues

A. Sfriso (✉) · C. Facca
Department of Environmental Sciences, University of Venice, Calle Larga Santa Marta 2137, 30123 Venice, Italy
e-mail: sfrisoad@unive.it

Introduction

In the last 50 years massive nutrient enrichments have affected worldwide coastal and transitional environments triggering abnormal and extensive growth of nuisance macroalgae (Rijstenbil & Haritonidis, 1993; Schramm & Nienhuis, 1996; Morand & Briand, 1996; Valiela et al., 1997; Raffaelli et al., 1998). Some genera of Chlorophyceae such as: *Ulva*, *Cladophora*, *Chaetomorpha* and Rhodophyceae such as *Gracilaria*, which are characterized by high growth rates and the ability to grow in dystrophic-hypertrophic environments, rapidly replaced the natural populations of macroalgae and seagrasses leading to strong environmental alterations (Valiela et al., 1997; Raffaelli et al., 1998). In the Mediterranean Sea the most extraordinary eutrophication processes were monitored in the lagoons of the northern Adriatic Sea: the Venice lagoon (Sfriso et al., 1988), the Po delta (Viaroli et al., 2001; Zaldìvar et al., 2003; Viaroli & Christian, 2004) as well as in the Étang du Prévost lagoon (Castel et al., 1996). In the Venice lagoon important changes were recorded from the '70s to the '90s, when in the central and northern basins macrophyte community structure underwent a sudden change, especially due to blooms of the chlorophycea *Ulva rigida* C. Ag. (Sfriso, 1987). That species, which is a not specialised green macroalga composed by two layers of cells provided with photosynthetic pigments able to uptake nutrients also in the organic form, is a significant example of those changes. In fact, it rapidly covered the lagoon bottoms with a biomass which increased exceptionally fast, reaching rates of 23–33% per day (Sfriso, 1995) and progressively reducing the seagrass beds. Up to that period only macrophyte taxonomic studies were available. The first quantitative study dates back to summer 1980. That year the biomass distribution of nuisance macroalgae was recorded considering ca. 2500 sampling sites, but only a map for the *Ulva* distribution in central lagoon was drawn and published (Solazzi et al., 1991; Curiel et al., 2004), therefore information on the total biomass covering the central lagoon and the other basins was not available. Successively, the macroalgal biomass, growth and production were studied during one whole year in many lagoon areas determining the production/biomass (P/B) and the gross/net production (GPP/NPP) ratios at different biomass levels. In June 1987, 1993 and 1998 the macroalgal biomass and production were recorded in the central lagoon only, allowing the estimation of the SC and the annual NPP and GPP (Sfriso et al., 2003). During the same years also phytoplankton distribution was recorded. Besides, other maps of some areas of the central lagoon dealing with the biomass distribution in 1989, 1990, 1992, 1994, 1996, 1998, were reported by Curiel et al. (2004). The first map of seagrasses dates back to 1990 (Caniglia et al., 1992) but it does not report quantitative results, as the map drawn in 2002 (Rismondo et al., 2003). In summer, 2003 the whole lagoon was monitored again. Biomass data of macrophytes (macroalgae and seagrasses) were recorded in 465 sites. Phytoplankton chlorophyll *a* and cell abundance were also monitored in 165 sites. The total standing crop (SC), net (NPP) and gross (GPP) production of the dominant macroalgae and the seagrasses *Cymodocea nodosa* (Ucria) Asherson, *Zostera marina* Linnaeus, *Nanozostera noltii* (Hornemann) Tomlinson *et* Posluzny, were recorded. A checklist for the macroalgae was also produced (Sfriso & La Rocca, 2005) whereas for the phytoplankton it is in progress.

This paper aims at providing and updating quantitative information on the primary producers of the Venice lagoon by estimating the coverage, the standing crop and the primary production of macroalgae and seagrasses, as well as the phytoplankton (chlorophyll *a*) distribution. The data on the macroalgal biomass recorded in 1980 in the whole lagoon have also been retrieved and new complete and inedited biomass maps drawn. They allow to evaluate the changes of primary producer communities which occurred in the lagoon of Venice in the last twenty years.

Methods

Study area

The Venice lagoon is a shallow water body located in the Northern Adriatic Sea which has a surface of ca. 549 km^2 and a depth of ca. 1.2 m. The lagoon is connected to the sea through three large (400–900 m) and deep (15–50 m) mouths which divide it into three hydrological basins separated by watersheds which shift according to tides and winds. Therefore, we refer to the three morphological basins: Northern, Central and Southern. Burano and Torcello tidal marshes mark the separation between the central and the northern basins and the deep Malamocco-Marghera artificial canal the one between the central and southern ones. Maps have been drawn considering the three basins separately.

Macroalgal and seagrass sampling

Sampling was carried out between June and August in 1980 and in 2003, during the highest biomass production. Sampling sites were 2500 in 1980 and 465 in 2003, but in 2003 seagrasses, phytoplankton and environmental parameters were also sampled. In 1980 macroalgae were recorded within 6 biomass ranges: 0.1–1, 1–5, 5–10, 10–15, 15–20, 20–25 kg fwt m^{-2}. In 2003, due to the biomass regression, the selected biomass ranges were: 0.01–0.1, 0.1–0.5, 0.5–1.0, 1.0–5.0, 5.0–10.0 kg fwt m^{-2}, but in the southern lagoon the highest range was only up to 1.0–2.0 kg fwt m^{-2}.

The three seagrass species (*C. nodosa*, *Z. marina* and *N. noltii*) were sampled separately. The biomass distribution was recorded according to 4 biomass ranges corresponding to a coverage of 0–25%, 25–50%, 50–75%, 75–100% and to a biomass ranging from 0–1.9 to 1.9–3.8, 3.8–5.6 and 5.6–7.5 kg fwt m^{-2} for *C. nodosa*, from 0–1.5 to 1.5–3.0, 3.0–4.5 and 4.5–6.3 kg fwt m^{-2} for *Z. marina* and from 0–1.4 to 1.4–2.7, 2.7–4.1 and 4.1–5.4 kg fwt m^{-2} for *N. noltii*. Those values were obtained by sampling the biomass and production during one year cycle in a grid of stations of the lagoon (Sfriso et al., 1993; Sfriso & Ghetti, 1998; Sfriso et al., 2004).

The calculation of the standing crop (SC) and the net (NPP) and gross (GPP) primary production are reported in Tables 1–3. The SC was obtained by summing the mean biomass calculated per each biomass range and lagoon surface. The accuracy for the biomass determination at each station was >95% (Sfriso et al., 1991; Sfriso & Ghetti, 1998), and ca. 90% when also the lagoon surface determination was considered.

The NPP and GPP were the results of the highest SC per biomass range multiplied by the

Table 1 Macroalgal standing crop and biomass production in the lagoon basins in 1980

Macroalgae, 1980									
Biomass range	Lagoon surface		Standing Crop			P/B	Annual biomass production		
kg fwt m^{-2}	km^2	%	mean ktonnes	min	max	annual	NPP ktonnes	GPP/NPP annual	GPP ktonnes
Northern basin									
20–25	0.9	3.0	20	18	23	1.6	36	6.7	242
15–20	1.6	5.3	28	24	32	1.6	51	6.7	345
10–15	5.2	17.2	65	52	78	1.6	124	6.7	833
5–10	4.3	14.4	33	22	43	2.0	87	6.7	581
1–5	9.5	31.4	28	9	47	3.5	166	6.7	1110
0.1–1	8.6	28.7	5	1	9	4.5	39	3.5	136
Total	30	100	179	126	232		503		3246
Central basin									
20–25	0.3	0.4	7	7	8	1.6	13	6.7	88
15–20	2.2	2.4	38	32	43	1.6	69	6.7	462
10–15	9.9	11.2	124	99	149	1.6	239	6.7	1598
5–10	19.3	21.8	145	97	193	2.0	387	6.7	2591
1–5	31.3	35.3	94	31	157	3.5	548	6.7	3674
0.1–1	25.6	28.9	14	3	26	4.5	115	3.5	403
Total	89	100	422	269	576		1371		8816
Southern basin									
20–25	0.4	0.5	9.7	9	11	1.6	17	6.7	116
15–20	0.6	0.7	10.8	9	12	1.6	20	6.7	132
10–15	3.3	4.0	41.6	33	50	1.6	80	6.7	535
5–10	6.9	8.4	51.8	35	69	2.0	138	6.7	926
1–5	35.5	42.9	106	35	177	3.5	621	6.7	4161
0.1–1	36.0	43.5	19.8	4	36	4.5	162	3.5	567
Total	83	100	240	125	356		1038		6437
Total (three basins)	202		841	520	1163		2912		18498

Table 2 Macroalgal standing crop and biomass production in the lagoon basins in 2003

Macroalgae, 2003									
Biomass range	Lagoon surface					P/B	Annual biomass production		
			mean	min	max		NPP	GPP/NPP	GPP
kg fwt m^{-2}	km^2	%	ktonnes			annual	ktonnes	annual	ktonnes
1–2	6.0	29.5	8.9	6.0	11.9	3.5	41.7	6.7	279
0.5–1	7.3	36.0	5.5	3.6	7.3	3.5	25.5	3.5	89
0.1-0.5	1.4	7.1	0.4	0.1	0.7	4.5	3.2	3.5	11
0.01–0.1	5.5	27.4	0.3	0.1	0.6	4.5	2.5	3.5	9
Total	20.2	100	15.1	9.8	20.5		73		388
Central basin									
1–5	1.2	2.9	3.61	1.2	6.0	3.5	21.1	6.7	141
0.5–1	6.8	16.3	5.10	3.4	6.8	3.5	23.8	3.5	83.4
0.1–0.5	1.4	3.3	0.42	0.1	0.7	4.5	3.1	3.5	11.0
0.01–0.1	32.4	77.5	1.78	0.3	3.2	4.5	14.6	3.5	51.0
Total	41.8	100	10.9	5.1	16.8		62.6		286
Southern basin									
5–10	1.23	1.4	9.2	6.1	12.3	2.0	25	6.7	165
1–5	7.27	8.3	21.8	7.3	36.4	3.5	127	6.7	853
0.5–1	25.7	29.2	19.3	12.9	25.7	3.5	90	3.5	315
0.1–0.5	38.7	43.9	11.6	3.9	19.3	4.5	87	3.5	304
0.01–0.1	15.2	17.2	0.8	0.2	1.5	4.5	7	3.5	24
Total	88.1	100	62.8	30.3	95.2		336		1661
Total (three basins)	150		89	45	132		471		2336

Table 3 Seagrass standing crop and biomass production in the lagoon basin in 2003

Cymodocea nodosa, 2003							
Biomass range	Lagoon surface					P/B	annual NPP
			mean	min	max		
kg fwt m^{-2}	km^2	%	ktonnes			annual	ktonnes
5.6–7.5	12.0	50.7	78.5	67	90	2.90	260
3.8–5.6	2.5	10.6	11.8	9	14	2.90	41
1.9–3.8	5.4	22.8	15.1	10	20	3.85	78
0.0–1.9	3.7	15.8	3.5	0	7	3.85	27
Total	23.6	100	109	87	131		406
Zostera marina, 2003							
4.5–6.3	11.0	42.1	59.3	49	69	3.30	228
3.0–4.5	5.10	19.6	19.1	15	23	3.30	76
1.5–3.0	3.10	11.9	7.0	5	9	3.30	31
0.0–1.5	6.88	26.4	5.2	0	10	3.30	34
Total	26.0	100	90	69	112		369
Nanozostera noltii, 2003							
4.1–5.4	0.61	9.8	2.9	2.5	3.3	3.30	11
2.7–4.1	0.71	11.4	2.4	1.9	2.9	3.30	9
1.4–2.7	0.94	15.1	1.9	1.3	2.5	3.30	8
0.0–1.4	3.96	63.7	2.7	0.0	5.3	3.30	18
Total	6.2	100	9.8	5.6	14.0		46
Total (three species)	55.9		209	162	257		821

P/B and GPP/NPP ratios, respectively. The P/B ratios were calculated according to Thorne-Miller & Harlin (1984) for the more abundant macroalgae of the lagoon. Those were sampled 2–4 times a month, for one entire year, in many sites of the lagoon since 1985–1986 (Sfriso et al., 1993; 2003). The production (P) was the sum of the positive biomass changes recorded during ca. 25–35 sampling dates in the period between the biomass peaks (B) monitored during two consecutive years. The biomass peak of *Ulva rigida* C. Ag. usually occurred in May-June and the biomass recorded at any sampling date was the mean of six 1 m^2 replicates. Sampling error, usually was lower than 5% (Sfriso et al., 1991) and only significant biomass changes between two sampling dates were considered (one-way ANOVA: $P < 0.05$). Similarly, two times a month samples, on an yearly basis, allowed to determine also *Z. marina* (Sfriso & Ghetti, 1998) and *C. nodosa* (Sfriso et al., 2004) P/B ratios, whereas *N. noltii* measurements are still in progress. Seagrass production was obtained by marking shoot leaf bundles and rhizome meristems according to Dennison (1990a, b) and summing plant growth during one year (Sfriso & Ghetti, 1998).

Ulva rigida GPP/NPP ratios were obtained by balancing a non-conservative element, such as phosphorus, in the macroalgal biomass, surface sediment and settled particulate matter in an area of the Venice lagoon watershed close to the Lido island, during different years (Sfriso & Marcomini, 1994). For seagrasses no GPP/NPP ratios were determined: the values quoted in this paper were estimated according to literature (Phillips & McRoy, 1980; Larkum et al., 1989; Philippart, 1994) by taking into consideration respiration processes, grazing and leaf and rhizome dead parts. The error estimation of NPP and GPP per area is ca. 10% and 20%, respectively, but the percentage increases to ca. 15% and 25% if errors related to map determination (lagoon surface) are also taken into account.

Phytoplankton determination

The distribution of phytoplankton was recorded as chlorophyll *a* (Chl *a*) and phaeophytin *a* (Phaeo *a*) concentrations in 165 sites where also nutrient and physico-chemical parameters were sampled. Waters samples were the results of 5–6 sub-samples of the entire water column collected with a Plexiglas bottle (height: ca. 1.5 m, i.d. 4 cm). Water aliquots of 250–1000 ml were filtered through GF/F Whatman glass filters (porosity 0.7 μm), placed in Petri disks, wrapped in aluminium foils and kept at –20°C till analyses. Pigment concentrations were spectrophotometrically determined according to Lorenzen (1967).

Results

Macroalgae

In 1980 macroalgae covered ca. 202 km^2 of the lagoon with a biomass higher than 0.1 kg fwt m^{-2} (Table 1, Fig. 1). Except for the zone close to Porto Marghera industrial area, the biomass covered the northern and central basins almost completely showing a mean SC of 5.9 and 4.8 kg fwt m^{-2}, respectively. The southern lagoon SC was, on average, only 2.9 kg fwt m^{-2}, but that basin was mainly colonised by seagrasses. Although the highest biomass range used to estimate the SC was 20–25 kg fwt m^{-2}, in some sampling sites, biomass peaked up to 35–40 kg fwt m^{-2}, due to accumulation by currents or bottom hollows. On the whole, the SC ranged from ca. 520 to 1163 ktonnes with a mean value of ca. 841 ktonnes, 422 of which were recorded in the central lagoon (Table 1). The annual NPP and GPP estimated by using P/B and GPP/NPP ratios were 2912 and 18498 ktonnes. In the northern lagoon the dominant species was *Valonia aegagropila* C. Ag. a globose chlorophyceaean which lives rolling on the bottom. This species accumulated in the bottom hollows reaching very high biomasses but, due to its shape, remained always submerged without affecting the water exchange and renewal. The central basin was mainly colonised by *Ulva rigida* C. Ag. The laminar thalli of that species trapped the oxygen bubbles produced in the photosynthetic processes during the sunny days and fluctuated in the water column hampering water exchanges and priming anoxia. The southern lagoon was mainly colonised by

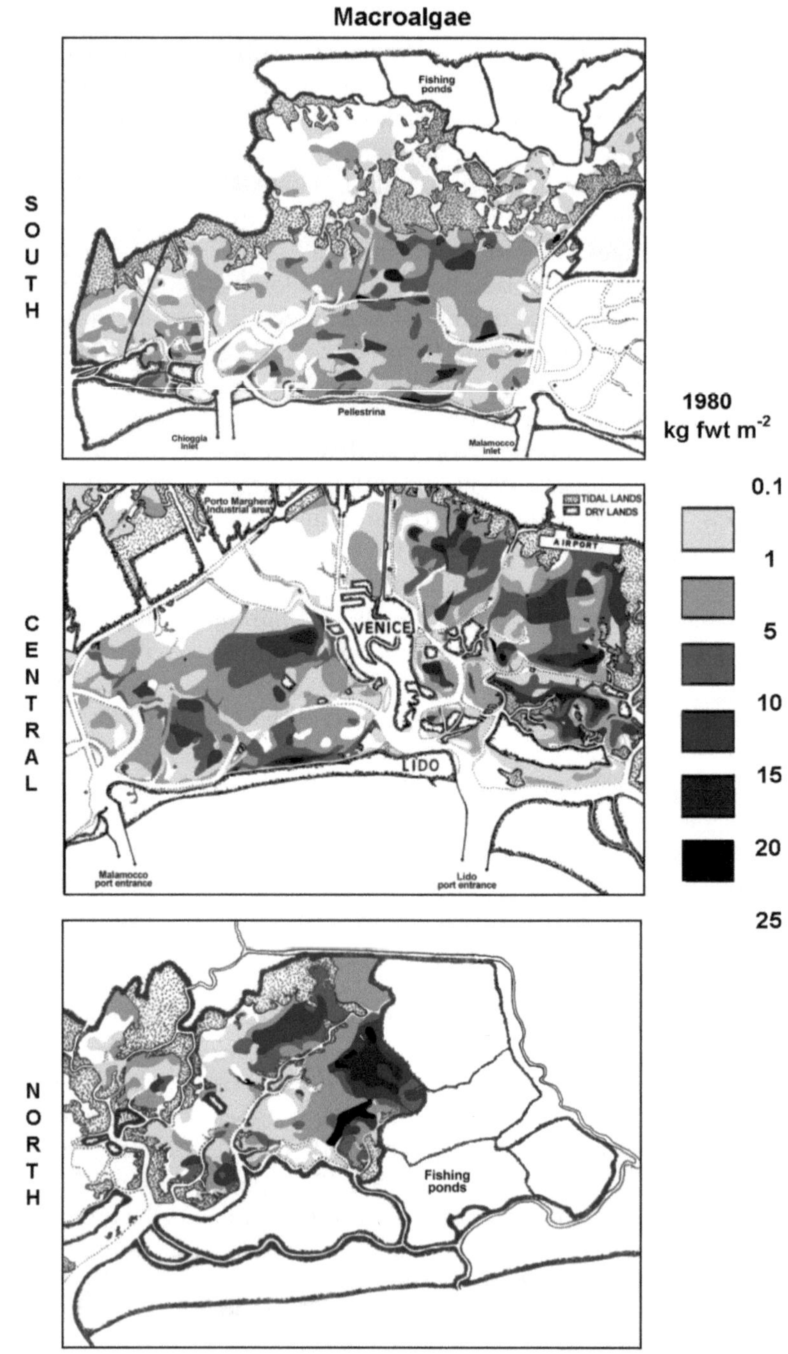

Fig. 1 1980 macroalgal distribution in the three lagoon sub-basins (South, Central, North). The biomass ranges are: 0.1–1, 1–5, 5–10, 10–15, 15–20 and 20–25 kg fwt m^{-2}. Pointed areas are tidal lands

seagrasses and the dominant macroalga was *Chaetomorpha linum* (O. F. Müller) Kützing a filamentous species which in the presence of high biomasses does not hamper water circulation.

In 2003, macroalgal biomass appeared strongly reduced when compared with 1980 results, especially the Central and Northern basins (Table 2, Fig. 2). The bottom coverage, starting with a

Fig. 2 2003 macroalgal distribution in the three lagoon basins (South, Central, North). The biomass ranges are: 0.01–0.1, 0.1–0.5, 0.5–1, 1–5 and 5–10 kg fwt m^{-2}. In the northern basin the highest biomass range is 1–2 kg fwt m^{-2}. Pointed areas are tidal lands

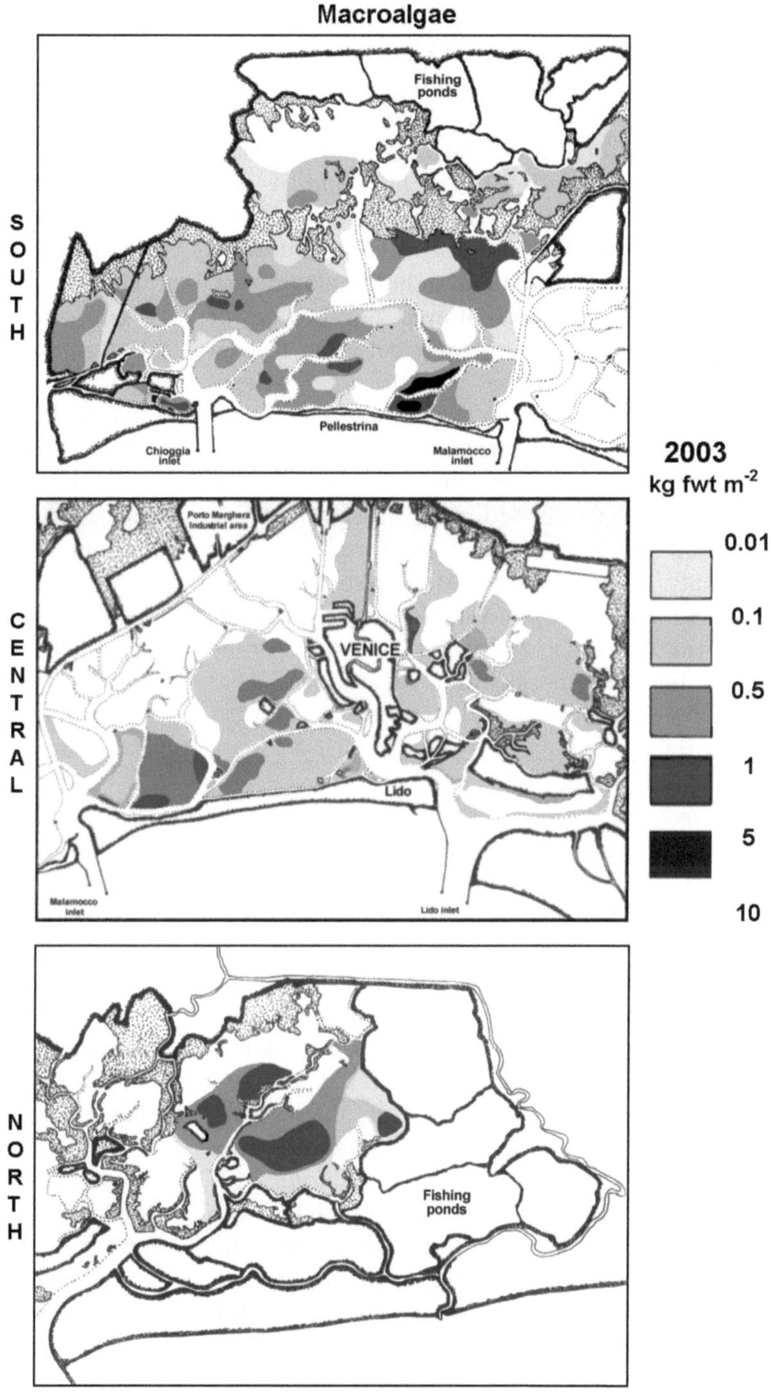

biomass >0.1 kg fwt m^{-2}, which is the minimum biomass sampled in 1980, was ca. 97 km^2. It increased to ca. 150 km^2 when also the lowest density (0.1–0.01 kg fwt m^{-2} biomass range) was considered.

On the whole, the SC ranged from 45 to 132 ktonnes with a mean value of 89 ktonnes (Table 2). The biomass showed the highest SC in the southern basin with ca. 63 ktonnes, a value 4–5 times as high as in the other basins. There, in an area facing the island of Pellestrina, biomass peaks were in the range 5–10 kg fwt m^{-2} (Fig. 2). The biomass was represented by a *C. linum* population which covered a *C. nodosa* bed.

In the central lagoon the maximum biomass was in the range 1–5 kg fwt m^{-2} and it was recorded on a *Z. marina* bed facing Lido island. In that case the biomass consisted in different species but *U. rigida* remained the dominant one.

The northern lagoon showed a much lower biomass with peaks ranging between 1 and 2 kg fwt m^{-2}, only. In that case the species absolutely dominant was *Vaucheria submarina* (Lyngbye) Berkeley. This filamentous species colonises very turbid areas, where other species are hampered, and contrasts sediment erosion forming dense beds which are strongly anchored to surface sediments. Filaments penetrate into mud bottoms favouring sedimentation. When beds are scattered, they form turfs which can rise ca. 5–15 cm over the surrounding bottoms devoid of vegetation.

Seagrasses

C. nodosa, Z. marina and *N. noltii* have been monitored keeping the three species separate (Table 3, Figs. 3–4). *Z. marina* and *C. nodosa* meadows had approximately the same size, with a surface of 26.0 and 23.6 km^2, respectively. *N. noltii* was restricted in a smaller area of 6.2 km^2, only. The highest SC (shoots and roots-rhizomes) was exhibited by *Cymodocea* with a mean of 109 ktonnes. At present, that species is the most abundant marine macrophyte in the Venice lagoon. The total seagrass SC was about twice as much (ca. 209 ktonnes) and *N. noltii* only contributed with 9.8 ktonnes. The NPP was 406 ktonnes for *C. nodosa* and 369 ktonnes for *Z. marina* whereas the preliminary NPP estimation for *N. noltii* was 46 ktonnes, but specific P/B ratios for that species are in progress.

The three species are frequently assembled showing mixed populations but generally they colonise very different environments. *C. nodosa* was mainly recorded in the southern basin in front of the Pellestrina island and in the areas close to the three sea inlets (Fig. 3) in coarse and well oxygenated sediments. Exceptionally, a new population was found also in the inner part of the northern basin in an area where this species has never been found before, which confirms the strong changes that have affected the lagoon during the last 15 years (Facca et al., 2002a, b; Sfriso et al., 2003; Pavoni et al., 2003; Secco et al., 2005).

Similarly, *Z. marina* was mainly recorded in the southern basin, particularly in inner areas where sediment are finer and nutrient-enriched (Fig. 4). That species, has recently colonised the central basin, near the Malamocco inlet, also covering the bottoms where clam-fishing activities are negligible or which in the past were colonised by macroalgae. *Z. marina* lacks in the northern basin except for the bottoms near the Lido inlet.

Until ten years ago *N. noltii* used to colonise large areas of the lagoon especially those near the salt marshes (Caniglia et al., 1992). During the 2003 it showed a marked regression and disappeared almost completely in the northern and central basins (Fig. 5). On the contrary, small patchy beds were found in the southern basin. Its maximum distribution was recorded near some salt marshes and close to Pellestrina island.

Phytoplankton

On the whole, phytoplankton showed a decreasing trend progressing from the mainland seaward and from the southern basin northward (Table 4, Fig. 6). The mean Chlorophyll *a* concentration was 2.29 ± 3.76 μg l^{-1} but the mean values decreased from 4.91 ± 8.47 μg l^{-1} in the northern basin to 2.15 ± 2.01 and 1.79 ± 2.76 μg l^{-1} in the central and southern ones. A 37 μg l^{-1} peak value was recorded in the northern basin close to the freshwater inputs of the rivers Dese and Siloncello. Relatively high concentrations were found in the central lagoon near the industrial zone (10.4 μg l^{-1}) and in the salt marshes of the southern basin (16.3 μg l^{-1}). Phaeopigments displayed the same trends and concentrations as chlorophyll. The total chlorophyll *a* (Chl *a* + Phaeo *a*) attained a peak of ca. 54 μg l^{-1}.

Discussion

At present, seagrasses are the main primary producers of the lagoon, whereas macroalgae

Fig. 3 Distribution of *Cymodocea nodosa* in the three lagoon basins. The biomass ranges are: 0–1.9, 1.9–3.8, 3.8–5.6 and 5.6–7.5 kg fwt m^{-2}. In the northern lagoon an arrow indicates a spot only near the fishing ponds. Pointed areas are tidal lands

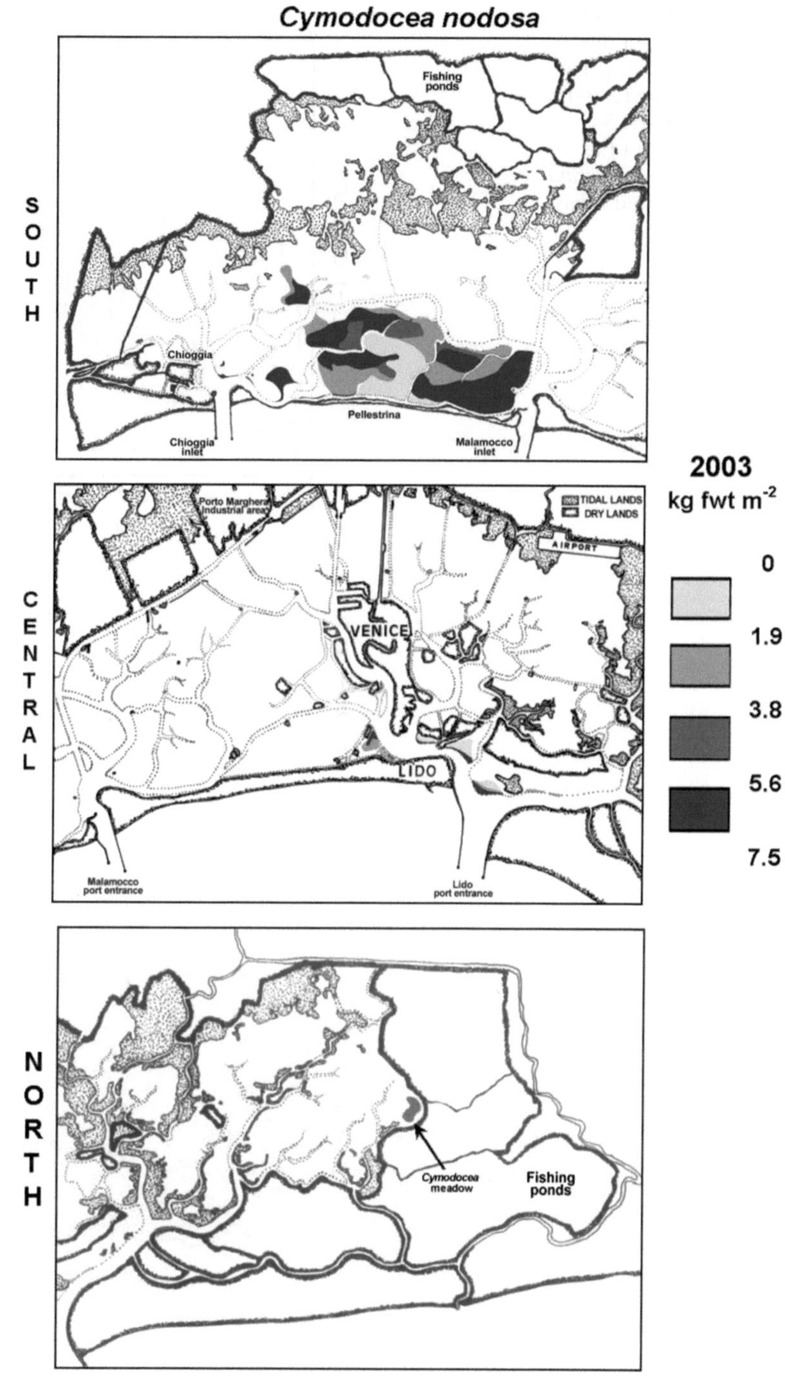

are retreating and phytoplankton appears to be low. In particular, macroalgae showed a remarkable difference between 1980 and 2003. In summer 2003 the SC, NPP and GPP were ca 10.6%, 16.2% and 12.7% of the values found in 1980 (Table 5) and these percentages decreased even to 2.6%, 4.6%, 3.4% if we consider the central basin only. Between the '70s and the '80s luxuriant beds of *U. rigida* and other ulvaceans replaced seagrass meadows especially in the

Fig. 4 Distribution of *Zostera marina* in the three lagoon basins. The biomass ranges are: 0–1.5, 1.5–3, 3–4.5 and 4.5–6.3 kg fwt m^{-2}. In the northern basin *Z. marina* is missing. Pointed areas are tidal lands

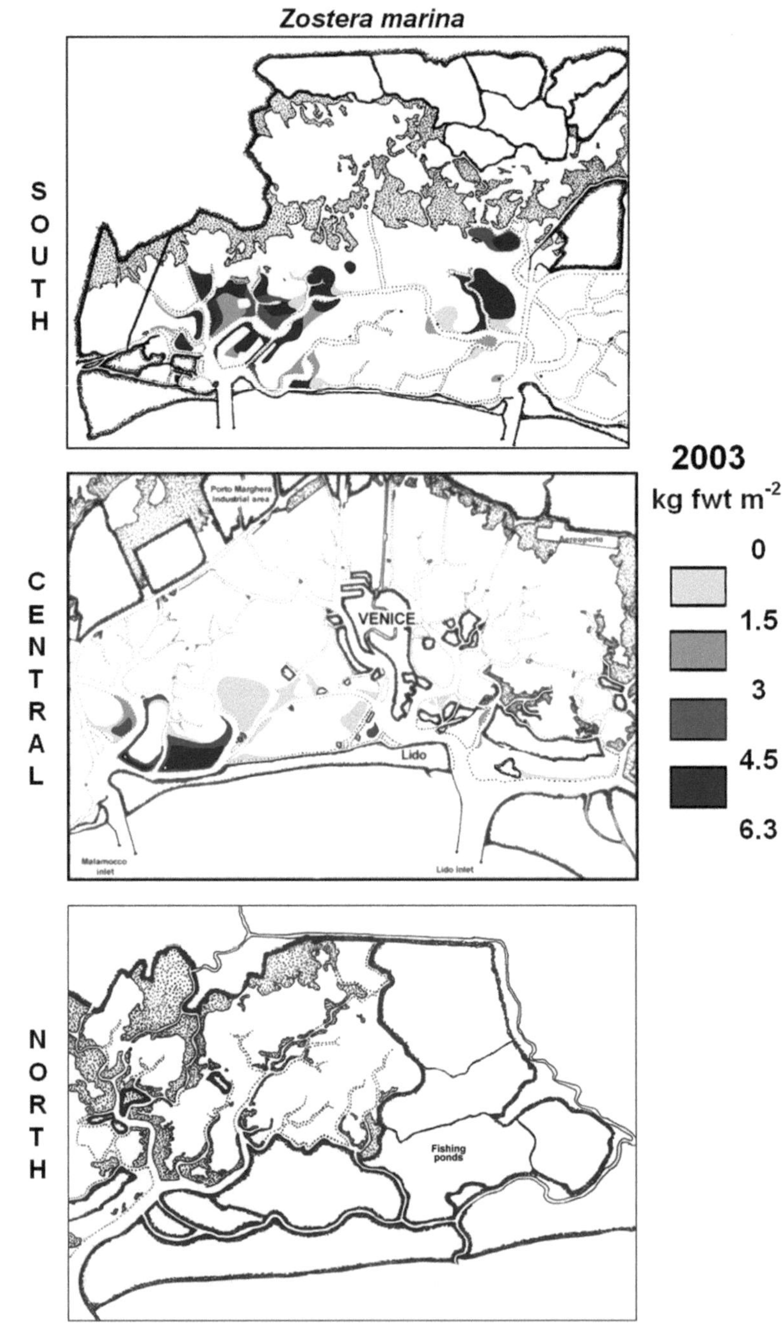

central basin. Due to the increase of nutrient availability, the ability of growing with floating thalli and the high relative-growth-rates, macroalgal species outcompeted seagrasses and caused abnormal blooms, the so-called "green tides". In late spring, macroalgal beds covered most of the lagoon till the increase of the water temperature triggered the biomass decomposition and anoxia. The consequence were the death of fish and macrofauna (Sfriso et al., 1988). Huge amounts of nutrients were released in surface sediments and in the water column triggering phytoplankton

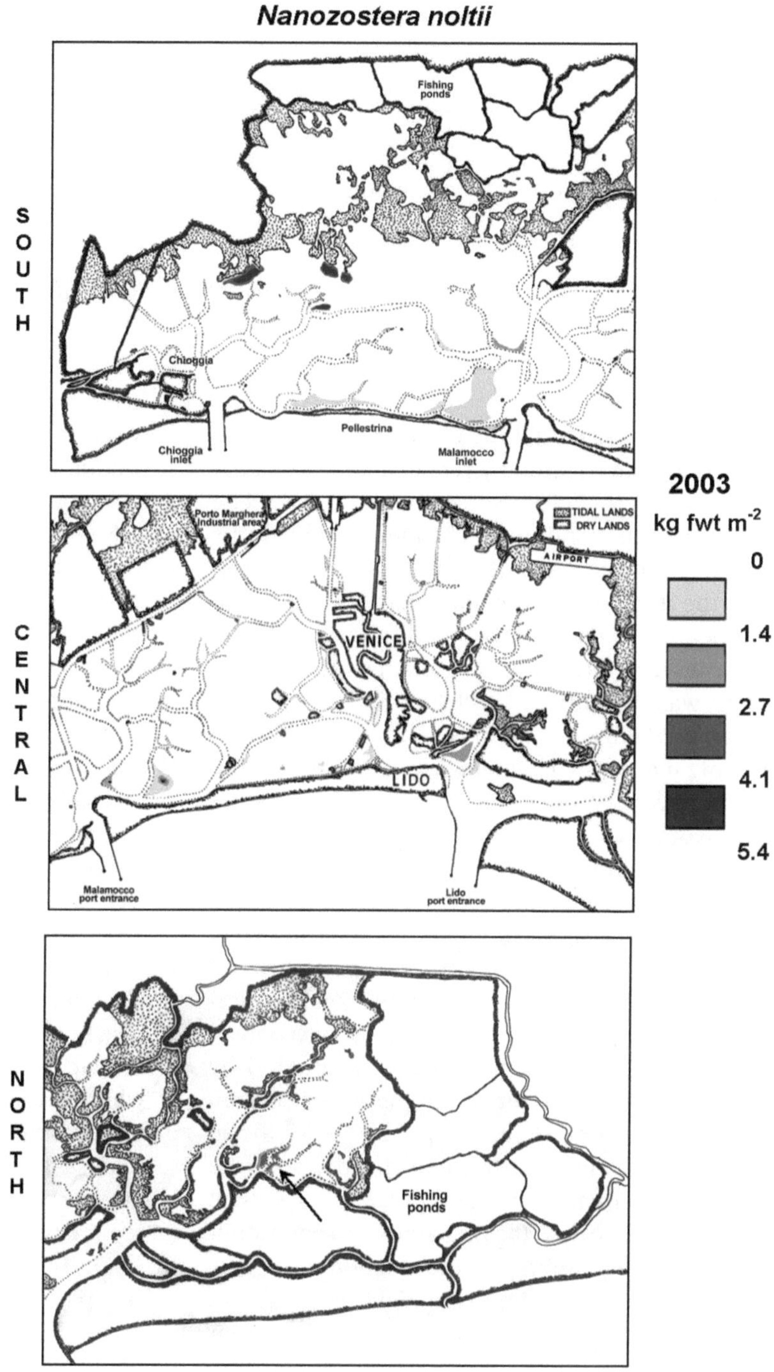

Fig. 5 Distribution of *Nanozostera noltii* in the three lagoon basins. The biomass ranges are: 0–1.4, 1.4–2.7, 2.7–4.1 and 4.1–5.4 kg fwt m^{-2} . In the northern lagoon some spots are placed only close to Lido inlet. In the northern lagoon an arrow indicates some spots placed near the fishing ponds. Pointed areas are tidal lands

blooms. After 1–2 months phytoplankton decreased, oxic conditions were restored and waters became very limpid again. Under those conditions macroalgae started to grow again till the winter pause. Then in spring the biomass increased very quickly covering all the lagoon and beginning a new annual cycle. Maps of the biomass distribution and production recorded in the summer of different years (Sfriso et al., 2003; Curiel et al., 2004) show that the macroalgae

Table 4 Chlorophyll *a* in the water column of the lagoon basins in 2003

Basin	N°		Chl *a* μg L^{-1}	Phaeo *a*	Chl *a* tot
North	19	mean ± std	4.91 ± 8.47	2.38 ± 3.95	7.29 ± 12.2
		max	37.0	16.7	53.7
Central	65	mean ± std	2.15 ± 2.01	1.92 ± 2.69	4.07 ± 4.16
		max	10.4	13.7	21.0
South	81	mean ± std	1.79 ± 2.76	1.87 ± 2.73	3.66 ± 4.95
		max	16.4	15.0	27.3
Total	165	mean ± std	2.29 ± 3.76	1.95 ± 2.86	4.24 ± 6.03
		max	37.0	16.7	53.7

dominance persisted till the early '90s. In 1993 the biomass appeared strongly reduced and 1998 data were similar to 2003 (Sfriso et al., 2003). The environmental conditions changed significantly, anoxia diminished and seagrasses started to colonize the areas where the macroalgal biomass had disappeared. Concurrently the Manila clam *Tapes philippinarum* Adams & Reeve, an exotic species which was recently introduced in the lagoon (Cesari & Pellizzato, 1985), spread on the bottom free of macrophytes changing completely the natural recovery of the lagoon. Clam-harvesting with hydraulic and mechanical gears disrupted the surface sediment texture and stopped seagrass recruitment because they were directly uprooted and because the finest sediment was continuously re-suspended, spread and settled. The increased turbidity reduced also phytoplankton blooms changing the composition of the microalgae communities which at present are dominated by benthic species re-suspended in the water column by clam-harvesting (Facca et al., 2002a, b; Sfriso et al., 2003), and contributed to depress the macroalgal growth. In the presence of high sedimentation rates, the macroalgae that dominated the vegetation of the lagoon changed. Species that need abundant light such as *V. aegagropila*, or are characterised by laminar thalli, such as *U. rigida*, disappeared or declined, whereas filamentous species such as the chrysophycea *Vaucheria submarina*, characterized both by thin reptant and erect filaments, grew in the surface sediments avoiding burial. At present, *Vaucheria* has colonised many areas with a biomass ranging from 0.5–1.2 kg m^{-2} fwt. That filamentous species colonises very turbid areas where other species are hampered and contrasts sediment erosion forming dense beds which are strongly anchored to the surface sediments.

Seagrasses, *C. nodosa* and *Z. marina* in particular, notwithstanding the anthropic impact, extended their beds (Rismondo et al., 2003) in the areas little influenced by clam-harvesting such as the southern basin and the bottoms close to the sea inlets. In the southern basin the beds of those species increased markedly in comparison with the maps drawn by Caniglia et al. (1992) notwithstanding the up-rooting action in some areas where clam-harvesting had been authorized in order to reduce the environmental effects of the free and uncontrolled clam-fishing (Orel et al., 2000). However, *C. nodosa* spreading depends almost exclusively on the growth and diffusion of its rhizomes because it is a subtropical species (Larkum et al., 1989) which hardly reproduces in the lagoon (Curiel et al., 1999; Sfriso et al., 2004), whereas the expansion of *Z. marina* occurs mainly by seedling (Sfriso & Ghetti, 1998). At present, *C. nodosa* is the most abundant macrophyte in the lagoon with a standing crop as high as the other seagrasses or seaweeds (Table 6).

On the contrary, *N. noltii*,, the species which in the past was dominant in the lagoon (Caniglia et al., 1992; Rismondo et al., 2003), has disappeared almost completely in the northern lagoon whereas near the Lido and Malamocco inlets and close to some salt marshes in the southern lagoon some beds have still remained. At present the biomass of *N. noltii* accounts for only 5% of the whole seagrass biomass. The increase of water turbidity, caused both by clam-harvesting and the works to rebuild or reinforce the embankments of

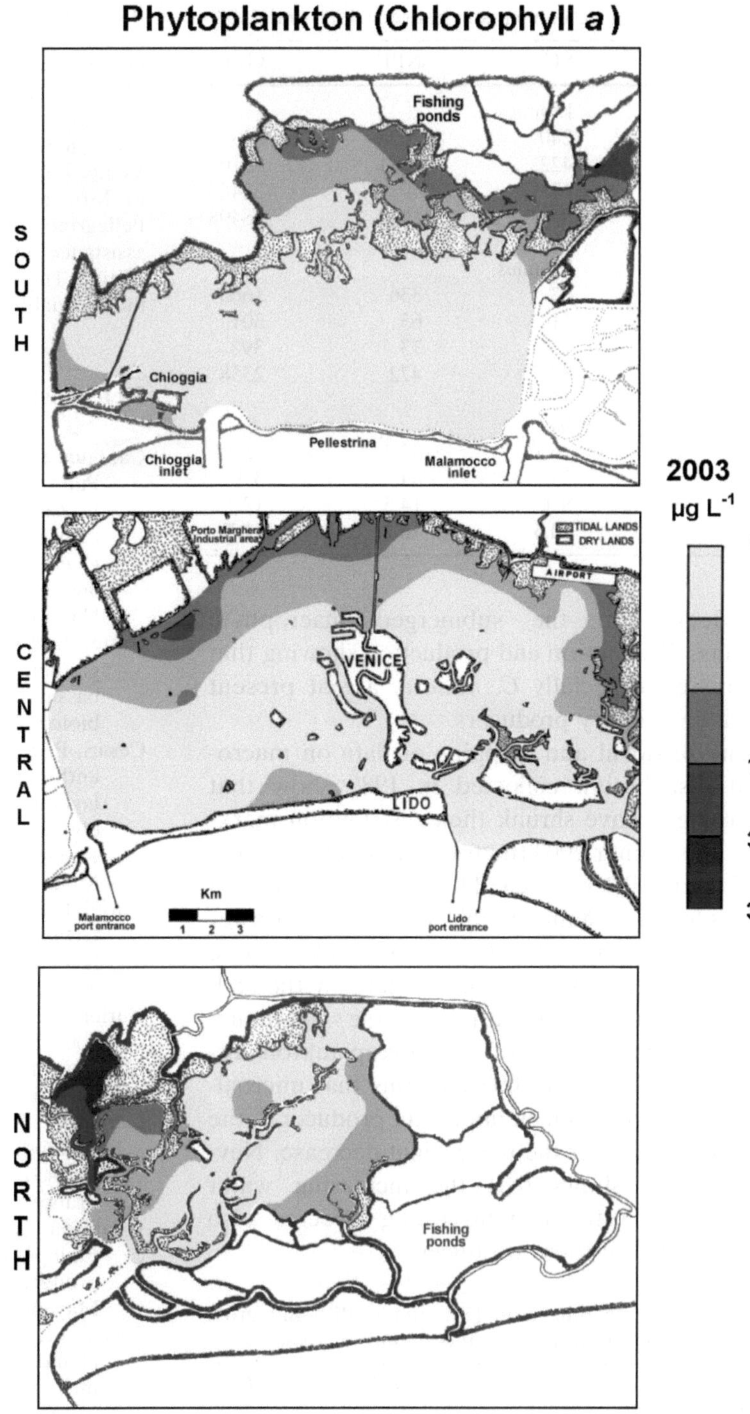

Fig. 6 Distribution of phytoplankton (Chl. *a*) in the three lagoon basins. Chlorophyll *a* ranges are: 0–3, 3–5, 5–10, 10–30 and 30–37 µg l^{-1}. Pointed areas are tidal lands

tidal marshes, as well as the employment of sandy sediments which are not suitable for the rooting of the species, seem the main cause of the reduction of this species in the lagoon.

Conclusions

Sampling surveys carried out in the whole Venice lagoon in summer 2003 allow to update the

Table 5 Macroalgae comparison between 1980 and 2003

Basin	SC	NPP	GPP
1980	ktonnes		
south	240	1038	6437
central	422	1371	8816
north	179	503	3246
Total	841	2912	18498
2003	ktonnes		
south	63	336	1665
central	11	63	301
north	15	73	392
Total	89	472	2358
2003/1980	%		
south	26.1	32.4	25.9
central	2.6	4.6	3.4
north	8.4	14.5	12.1
Total	**10.6**	**16.2**	**12.7**

knowledge on the submerged macrophyte biomass distribution and production showing that seagrasses, especially *C. nodosa*, are at present the main primary producers.

The retrieval and mapping of data on macroalgal distribution collected in 1980 show that macroalgae have shrunk their distribution markedly. In summer 2003 the total SC and the annual NPP and GPP were ca. 11, 16 and 13% of the values found in summer 1980, respectively. Lower values can be found (2.6, 4.6 and 3.4%) if one considers the central lagoon, where in the '80s nuisance macroalgae attained the maximum development. A map of the phytoplankton distribution in summer 2003 confirms that microalgae have never been an important producer in the lagoon, even after the macroalgal decrease. Nevertheless it shows that the increasing water turbidity due to clam-harvesting affected also the presence of those producers.

Table 6 total standing crop (SC) net (NPP) and gross (GPP) production in the lagoon of Venice in 2003

Basin	SC	NPP	GPP
2003			
	ktonnes		
Macroalgae	89	472	2336
C. nodosa	109	406	812
Z. marina	90	369	738
N. noltii	9.8	46	92
Total	298	1293	3978

Acknowledgements The authors thank the "Study Centre for the Application of Energetic Resources" (CSARE) that in part funded the macroalgal mapping in summer 1980. Macrophyte and phytoplankton measurements carried out in summer 2003 were supported by the "Consortium for Coordination of Research Activities and concerning the Venice Lagoon System" (CoRiLa).The authors are grateful to Mrs. Sonia Ceoldo, Mr. Fabio Cavolo, dr. Nicola Pellegrino and dr. Massimo Nicosia for the technical assistance and to dr. Orietta Zucchetta for the English editing. The authors thank also the anonymous referees for their valuable suggestions.

References

Caniglia, G., S. Borella, D. Curiel, P. Nascimbeni, A. F. Paloschi, A. Rismondo, F. Scarton, D. Tagliapietra & L. Zanella, 1992. Distribution of the marine phanerogams [*Zostera marina* L., *Zostera noltii* Hornem., *Cymodocea nodosa* (Ucria) Asch.] in the Venice lagoon (in italian). Lavori- Società Veneta di Scienze Naturali 17: 137–150.

Castel, J., P. Caumette & R. Herbert, 1996. Eutrophication gradients in coastal lagoons as exemplified by the bassin d'Arcachon and the Etang du Prévost. Hydrobiologia 329: 9–28.

Cesari, P. & M. Pellizzato, 1985. Voluntary or accidental anthropic introduction of molluscs in the Venice lagoon. *Saccostrea commercialis* (Iredale and Roughely, 1933) and *Tapes philippinarum* (Adams and Reeve, 1850) acclimatization (in italian). Bollettino Malacologico 21: 237–274.

Curiel, D., A. Rismondo & G. Bellemo, 1999. Flowering of *Cymodocea nodosa* (Ucria) Ascherson in the Venice lagoon (in Italian). Lavori–Società Veneta di Scienze Naturali 24: 133–134.

Curiel, D., A. Rismondo, G. Bellemo & M. Marzocchi, 2004. Macroalgal biomass and species variations in the Lagoon of Venice (Northern Adriatic Sea, Italy). Scientia Marina 68: 1981–1998.

Dennison, W. C. 1990a. Seagrass research methods. In Phillips, R. C. & C. P. McRoy (eds), Leaf production. Unesco, Imprimerie de la Manutention, Mayenne, France, 77–79.

Dennison, W. C. 1990b. Seagrass research methods. In Phillips, R. C. & C. P. McRoy (eds), Rhizome-root production. Unesco, Imprimerie de la Manutention, Mayenne, France, 81–82.

Facca, C., A. Sfriso & G. Socal, 2002a. Changes in abundance and composition of phytoplankton and microphytobenthos due to increased sediment fluxes in the Venice lagoon, Italy. Estuarine, Coastal and Shelf Science 54: 773–792.

Facca, C., A. Sfriso & G. Socal, 2002b. Temporal and spatial distribution of diatoms in the surface sediment of the central part of the Venice Lagoon. Botanica Marina 45: 170–183.

Larkum, A. W. D., A. J. McComb & S. A. Shepherd, 1989. Biology of Seagrasses. Elsevier, Amsterdam.

Lorenzen C. J., 1967. Determination of chlorophyll in sea water. UNESCO Technical Paper on Marine Sciences, 35–120.

Morand, P. & A. Briand, 1996. Excessive growth of macroalgae: A symptom of environmental disturbance. Botanica Marina 39: 491–516.

Orel, G., V. Boatto, A. Sfriso & M. Pellizzato, 2000. Fishing sources management plan for the lagoons of the Venice Province (in italian). Province of Venice, Venice.

Pavoni, B., M. Caliceti L. Sperni & A. Sfriso, 2003. Organic micropollutants (PAHs, PCBs, Pesticides) in seaweeds of the lagoon of Venice. Oceanologica Acta 26: 585–596.

Phillips, R. C. & C. P. McRoy, 1980. Handbook of Seagrass Biology: An Ecosystem Perspective. Garland STPM Press, New York.

Philippart, C. J. M., 1994. Eurtrophication as a possible cause of decline in the seagrass *Zostera noltii* of the Dutch Wadden Sea. Thesis Wageningen, Koninklijke bibliotheek, Den Haag.

Raffaelli, D. G., J. A. Raven & L. J. Poole, 1998. Ecological impact of green macroalgal blooms. Oceanography and Marine Biology Annual Review 36: 97–125.

Rismondo, A., D. Curiel, F. Scarton, D. Mion & G. Caniglia, 2003. A new seagrass map for the Venice lagoon. In Özhan, E. (ed.), Proceedings of the Sixth International Conference on the Mediterranean Coastal Environment, MEDCOAST 03, Ravenna, Italy.

Rijstenbil, J. W. & S. Haritonidis, 1993. Macroalgae, eutrophication and trace metal cycling in estuaries and lagoons. Symposium Proceedings Sub Group III, Action COST–48 (Aquatic Primary Biomass-Marine Macroalgae). Commission of the European Communities. BRIDGE, Brussels.

Schramm, W. & P. N. Nienhuis, 1996. Marine Benthic Vegetation, Ecological Studies. Springer Verlag, Heidelberg.

Secco, T., F. Pellizzato, A. Sfriso & B. Pavoni, 2005. The changing state of contamination in the lagoon of Venice. Part 1. Organic pollutants. Chemosphere 58: 279–290.

Sfriso A., 1987. Flora and vertical distribution of macroalgae in the lagoon of Venice: a comparison with previous studies. Giornale Botanico Italiano 121: 69–85.

Sfriso, A., B. Pavoni, A. Marcomini & A. A. Orio, 1988. Annual variation of nutrients in the lagoon of Venice. Marine Pollution Bulletin 19: 54–60.

Sfriso, A., S. Raccanelli, B. Pavoni & A. Marcomini, 1991. Sampling strategies for measuring macroalgal biomass in the shallow waters of the Venice lagoon. Environmental Technology 12: 263–269.

Sfriso, A., A. Marcomini, B. Pavoni & A. A. Orio, 1993. Species composition, biomass and net primary production in coastal shallow waters: the Venice lagoon. Bioresource Technology 44: 235–250.

Sfriso, A. & A. Marcomini, 1994. Gross primary production and nutrient behaviours in shallow lagoon waters. Bioresource Technology 47: 59–66.

Sfriso A., 1995. Temporal and spatial responses of *Ulva rigida* C. Ag. growth to environmental and tissue concentrations of nutrients in the lagoon of Venice. Botanica Marina 38: 557–573.

Sfriso, A. & P. F. Ghetti, 1998. Seasonal variation in the biomass, morphometric parameters and production of rhizophytes in the lagoon of Venice. Aquatic Botany 61: 207–223.

Sfriso, A., C. Facca & P. F. Ghetti, 2003. Temporal and spatial changes of macroalgae and phytoplankton in shallow coastal areas: the Venice lagoon as a study case. Marine Environmental Research 56: 617–636.

Sfriso, A., C. Facca & S. Ceoldo, 2004. Growth and production of *Cymodocea nodosa* (Ucria) Ascherson in the Venice lagoon. In Campostrini, P. (ed.), Scientific Research and Safeguarding of Venice. CoRiLa. Research Programme 2001–2003. 2002 Results. Venice, 229–236.

Sfriso, A. & B. La Rocca, 2005. Macroalgal taxa updating in littorals and shallow bottoms of the Venice lagoon (in italian). Lavori–Società Veneta di Scienze Naturali 30: 45–58.

Solazzi, A., G. Orel, E. Chiozzotto, M. Scattolin, D. Curiel, F. Grim, F. Aleffi, D. Del Piero & P. Vatta, 1991. The seaweeds of the Venice lagoon (in italian). Arsenale Editrice, Venice.

Thorne-Miller, B. & M. M. Harlin, 1984. The production of *Zostera marina* L. and other submerged macrophytes in a coastal lagoon in Rhode Island. U.S.A. Botanica Marina 27: 539–546.

Valiela, I., J. McLelland, J. Hauxwell, P. J. Behr, D. Hersh & K. Foreman, 1997. Macroalgal blooms in shallow estuaries: controls and ecophysiological and ecosystem consequences. Limnology and Oceanography 42: 1105–1118.

Viaroli, P., R. Azzoni, M. Bartoli, G., Giordani & L. Tajè, 2001. Evolution of the Trophic Conditions and Dystrophic Outbreaks in the Sacca di Goro Lagoon (Northern Adriatic Sea). In Faranda, F. M., L. Guglielmo & G. Spezie (eds), Mediterranean Ecosystems. Structure and Processes. Springer Verlag Italia, Milano, 467–475.

Viaroli, P. & R. R. Christian, 2004. Description of trophic status, hyperautotrophy and dystrophy of a coastal lagoon through a potential oxygen production and consumption index-TOSI: Trophic Oxygen Status Index. Ecological Indicators 3: 237–250.

Zaldívar, J. M., E. Cattaneo, M. Plus, C. N. Murray, G. Giordani & P. Viaroli, 2003. Long-term simulation of main biogeochemical events in a coastal lagoon: Sacca di Goro (Northern Adriatic Coast, Italy). Continental Shelf Research 23: 1847–1875.

Hydrobiologia (2007) 577:87–94
DOI 10.1007/s10750-006-0419-2

Growth and survival of the invasive alga, *Caulerpa taxifolia*, in different salinities and temperatures: implications for coastal lake management

E. J. West · R. J. West

Abstract The alga *Caulerpa taxifolia* is an invasive pest species in many parts of the world and has recently become established in several estuaries in south eastern Australia. A major infestation has occurred in Lake Conjola, an intermittently open and closed coastal lagoon in southern NSW. Short term (1 week) laboratory experiments were carried out to investigate growth and survival of fragments of *C. taxifolia* collected from this outbreak, under a range of salinities (15–30 ppt) and water temperatures (15–30°C). Fronds, stolons and thalli of the alga all displayed similar responses. Many of the algal fragments doubled in size over the week and a maximum growth rate of 174 mm/week was recorded. Fragments showed good growth (>20 mm/week) at salinities >20 ppt and temperatures >20°C. Almost total mortality occurred at salinities lower than 20 ppt and temperatures less than 20°C. Historical records of water quality demonstrate that prior to entrance manipulation in 2001, salinities in Lake Conjola had often dropped to below 17 ppt for extended periods (up to 2 years). This suggests that management of the alga may be improved if the lake was allowed to undergo its normal cycles of opening and closing to the ocean, and that entrance manipulation may be one factor that has influenced the success of this invasive species.

Keywords *Caulerpa taxifolia* · Salinity · Temperature · Growth · Coastal lagoon · Management

Guest editors: P. Viaroli, P. Lasserre & P. Campostrini
Lagoons and Coastal Wetlands in the Global Change Context: Impacts and Management Issues

E. J. West · R. J. West
Institute for Conservation Biology, University of Wollongong, Wollongong, NSW 2522, Australia

Present Address:
E. J. West (✉)
School of Environmental and Applied Sciences, Griffith University, PMB 50 Gold Coast Mail Centre, Queensland 9726, Australia
e-mail: E.West@griffith.edu.au

Introduction

Caulerpa taxifolia is a fast growing green alga, native to tropical waters of the Indian, Pacific and Atlantic Oceans (Phillips and Price, 2002). *C. taxifolia* has been popularly dubbed the 'killer algae' due to its success as an introduced noxious weed in several temperate locations, including Europe, USA and Australia (Jousson et al., 2000). Its high profile as an invasive species arises from its introduction to, and now widespread occurrence throughout, the Mediterranean Sea and its ability to out-compete important native species

(Boudouresque et al., 1995; Ribera et al., 1996; Ceccherelli and Cinelli, 1999b; Ceccherelli et al., 2002). The first reported introduction of *C. taxifolia* to non-tropical waters was a one square metre patch in the Mediterranean, off the Monaco coast, in 1984 (Meinesz et al., 1993). The species has now been reported widely throughout the Mediterranean at depths ranging from a few metres to around 100 m in clear water (Belsher and Meinesz, 1995). By 2000, *C. taxifolia* had infested approximately 131 km^2 composed of 103 independent colonies along 191 km of coastline in six countries: Spain, France, Monaco, Italy, Croatia, and Tunisia (Meinesz, 2002). In Australia, *C. taxifolia* is native to sheltered tropical waters throughout the north, ranging as far south as 28°15′ S on the eastern coastline (Phillips and Price, 2002). In April 2000, *C. taxifolia* was discovered in the temperate waters of New South Wales (NSW) and is now present in nine waterways.

C. taxifolia, along with most other species of siphonous green algae, can spread quickly and reach high densities because they grow rapidly relative to the seagrasses they replace (Ceccherelli and Cinelli, 1999a; Vroom and Smith, 2001; Ceccherelli et al., 2002). *C. taxifolia* is able to reproduce both sexually and asexually (Silva, 2002). However, sexual reproduction has not yet been observed in invasive populations (Zuljevic and Antolic, 2000). The success of *C. taxifolia* as an invasive species has therefore been largely attributed to its ability to reproduce asexually via vegetative fragments (Ashton and Mitchell, 1989; Ceccherelli and Cinelli, 1999b; Smith and Walters, 1999; Boudouresque and Verlaque, 2002). Very small fragments of *C. taxifolia*, some as small as 1 cm, can survive and have been successfully grown (Smith and Walters, 1999). Fragments can be produced by natural disturbances, such as storms, or created by human activities, such as boat propellers or anchors (Meinesz, 2002). The ability of *C. taxifolia* to fragment readily, successfully re-attach, to form new colonies, and to grow quickly, are key factors contributing to the rapid establishment and vegetative spread of the alga (Smith and Walters, 1999; Vroom and Smith, 2001).

In order to assist in the management of this pest alga, it is important to understand the environmental conditions that favour establishment and growth, particularly of fragments, which are the prime mode of spreading. Smith and Walters, (1999) examined growth of native *C. taxifolia* in laboratory experiments and found differences depending on the size of fragments and the position within the plant. Laboratory experiments carried out on the invasive *C. taxifolia* in the Mediterranean, have shown that *C. taxifolia* had little growth at temperatures between 10 and 15°C (Gayol et al., 1995; Komatsu et al., 1997). However, there is doubt about the similarity in the strains of *C. taxifolia* throughout the world (Boudouresque et al., 1995; Meinesz and Bouderesque, 1997; Schaffelke et al., 2002), and it is therefore important to understand the environmental conditions that favour growth of the invasive strain of *C. taxifolia* in the south east Australian region. Currently, there are no published experimental investigations on the growth and survival of *C. taxifolia* in this region. Here we examine the effects of salinity and temperature on the growth and survivorship of frond, stolon and thallus fragments of *C. taxifolia*. The findings are used to discuss the implications for the management of Lake Conjola, a medium sized coastal lagoon in this region, which is heavily infested with *C. taxifolia*.

Materials and methods

To examine the effects of temperature and salinity on the growth and survival of *C. taxifolia* fragments, experiments were done under controlled conditions in a growth cabinet (Thermoline Refrigerated Incubator), with controlled temperature (0–35°C) and lighting (neon tubes with PAR ~100 $\mu mol\ m^{-2}\ s^{-1}$). A 12-h day/night cycle was maintained during each experiment. *C. taxifolia* fragments and seawater were collected from Lake Conjola, NSW, Australia (34°05′ S, 151°08′ E), and transported to the laboratory at the beginning of each experiment so that fresh material was used each time. Lake

Conjola, which is a coastal barrier lagoon located approximately 210 km south of Sydney, is one of many lakes in this region that are characterised by entrance channels which open and close to the ocean depending on climatic conditions (West et al., 1985). *C. taxifolia* was first recorded at Lake Conjola in 2000, and the lake is now the most severely infested location in south east Australia.

A short pilot study was done to select the appropriate salinities and replication for the main laboratory experiments. In the pilot study, five replicate fragments were placed in the growth cabinet at 25°C at four salinities (10, 15, 25 and 30 ppt). These salinities were chosen after examination of published historical records of salinity recorded for Lake Conjola (SCC, 2003). It was found that the five replicates at 10 and 15 ppt were all dead after 1 week but that all replicates at 25 and 30 ppt grew both fronds and stolons during this short period. On the basis of these results, a range of salinities between 15 and 30 ppt was chosen for the main laboratory experiments.

For the main study, growth and survival of *C. taxifolia* fragments were assessed in a series of four experiments in the growth cabinet (described above). Each experiment was carried out over a week and at one of the following temperatures: 15, 20, 25 or 30°C. In each of the four experiments, 54 plastic containers were filled with 1 l of water, into which a fragment of *C. taxifolia* was placed. The water was a mixture of seawater from Lake Conjola and distilled water, adjusted to one of six salinities, namely: 15, 17.5, 20, 22.5, 25 and 30 ppt.

Growth and/or mortality were compared among three different fragment types, stolon (fragment of stolon only), frond (fragment of frond only) and thallus (fragment of stolon with one frond). There were three replicate fragments (<200 mm in length and <0.05 g dry weight) for each combination of fragment type and salinity. Single fragments were cut from relatively large and healthy plants kept at 30 ppt, immediately placed into open vessels containing water at the range of salinities listed above, and then randomly positioned in the growth cabinet. After one week, new fronds and stolons were counted and measured (length, mm) for each fragment. Total new growth (mm) was calculated by adding the new frond and stolon growth together for each fragment. This new growth was a combination of cell division and elongation. The catchment of Conjola Lake is largely cleared, and the estuarine waters of the lake have levels of dissolved nutrients (SCC, 2003) and inorganic carbon that are unlikely to be growth limiting for *C. taxifolia*. Since very small fragments of *C. taxifolia* were used in each container, and growth was measured after only 1 week, the effect of dilution on growth and mortality was considered to be primarily a response to salinity change. Mortality of fragments was measured as the proportion of the fragment that went a clear/white colour during the course of the experiments (% bleached).

Two-way ANOVAs were done to test for significant differences in total new growth [$\log_{10}(x + 1)$ transformed] and/or mortality [% bleached, $\arcsin(x/100)$ transformed] among salinities (fixed factor) and fragment type (fixed factor). Data were tested for homogeneity of variances using Cochran's test. Where all replicates of a treatment showed no growth or died, these data were excluded from the analyses to improve normality. Where some deviations from normality remained in the data, ANOVA was still adopted as this test is quite robust under such circumstances (Zar, 1984). Where significant differences were found, Tukey's HSD test was used for means comparison. Statistical comparisons were not made among temperatures because experiments were done at different times with different batches of *C. taxifolia* and therefore confounded by psuedoreplication (Hurlbert, 1984). Long term salinity and temperature records for Lake Conjola were obtained from the Shoalhaven City Council (SCC, 2003).

Results

Fragments of *C. taxifolia* were grown in the laboratory for periods of 1 week, during which time they had rapid growth rates at salinities >22.5 ppt and temperatures >15°C. The most growth displayed by a single fragment was 174 mm, and many fragments doubled in size during the 1-week study period. All fragments types displayed growth under suitable conditions.

In the majority of cases, fragments produced new stolons and fronds, rather than simply displaying repair and expansion of the damaged areas.

There was no growth of *C. taxifolia* at salinities of 20 ppt or less, regardless of temperature or fragment type (Fig. 1a). Fragments displayed some growth at salinities of 22.5 ppt and above, regardless of temperature, except for temperatures of 15°C where growth rate was negligible or small (Fig. 1a). At 20°C, highest growth rates were at 25 and 30 ppt; at 25°C, there was similar growth at the salinities greater than 20 ppt; while, at 30°C, the amount of growth was significantly higher at 30 ppt (Fig. 1a, Table 1). At salinities of

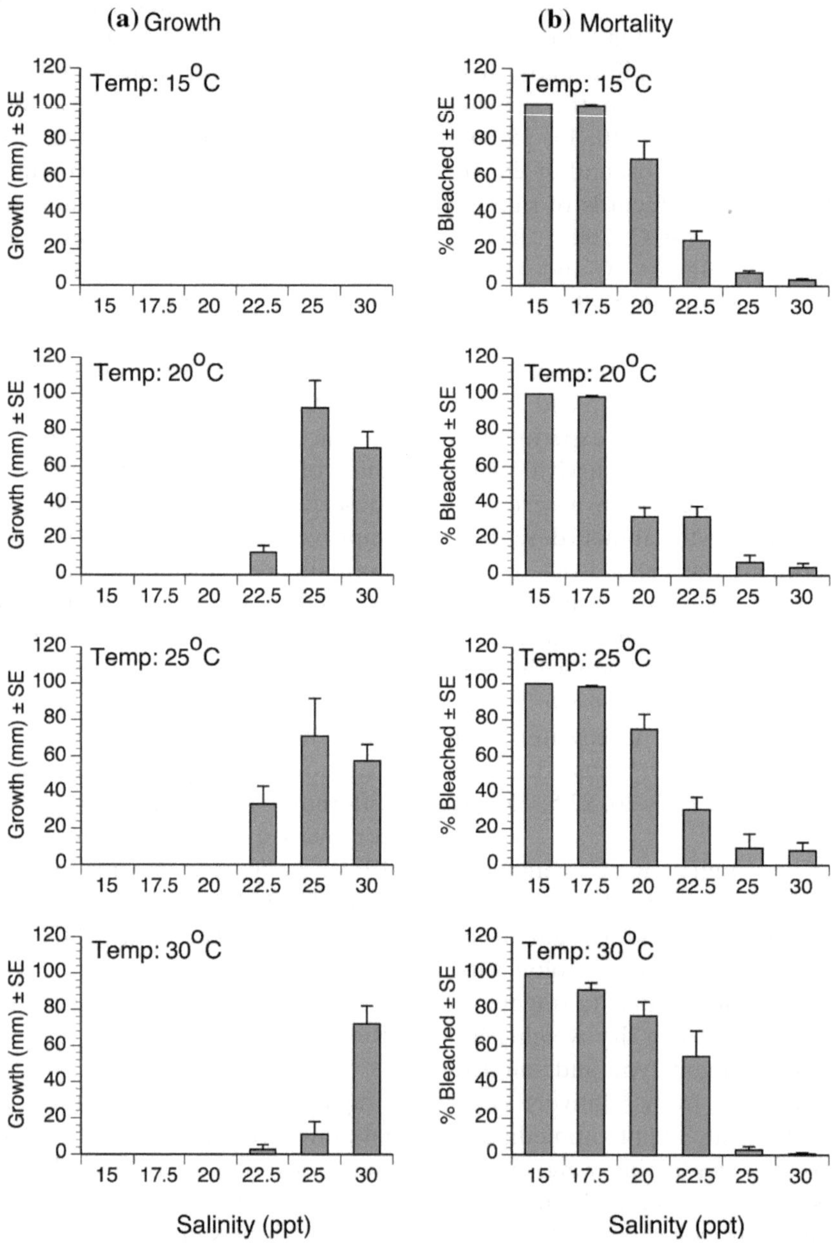

Fig. 1 (**a**) Growth (mm) and (**b**) Mortality (% bleached) of *C. taxifolia* fragments kept at different salinities and temperatures under laboratory conditions for 1 week. Data has been pooled over three fragment types (stolons, fronds and thallus). Mean and standard error shown (n = 9)

Table 1 Results of ANOVAs to test for differences in new growth (mm) of *C. taxifolia* fragments grown in laboratory conditions at four separate temperatures

Temperature		15°C		20°C		25°C		30°C	
Source	df	MS	F	MS	F	MS	F	MS	F
Salinity (S)	2	n.a.	n.a.	3.23	16.1***	0.37	0.7^{ns}	6.88	23.0***
Fragment (F)	2			0.04	0.2^{ns}	0.39	0.7^{ns}	0.24	0.8^{ns}
S × F	4			0.06	0.3^{ns}	1.24	2.3^{ns}	0.04	0.2^{ns}
Residual	18			0.20		0.55		0.30	
Cochran's		ns		ns		ns		ns	
Tukey's HSD				22.5 < 25 = 30 ppt		22.5 = 25 = 30 ppt		22.5 = 25 < 30 ppt	

^{ns} not significant; *<0.05; **<0.01; ***<0.001

Tukey's HSD test has been used to indicate significantly different means. Notes: (1) data for salinities of 15, 17.5 and 20 ppt were excluded from the analyses as there was no growth; (2) growth data has been log $10(x + 1)$ transformed

15 and 17.5 ppt, there was nearly 100% mortality of all *C. taxifolia* fragments, regardless of temperature (Fig. 1b). Mortality decreased at salinities 20 and 22.5 ppt, and the lowest mortality of fragments was at 25 and 30 ppt regardless of temperature (Fig. 1b; Table 2).

Long-term data for the salinity and temperature for Lake Conjola have been summarised for a representative site, located near the *C. taxifolia* outbreak, namely Roberts Point (Fig. 2).

Discussion

In general, *C. taxifolia* fragments grew rapidly in the laboratory at temperatures above 15°C and at salinities above 22.5 ppt. For example, several individual fragments doubled in size in 1 week. Successful growth of *C. taxifolia* fragments has also been documented in laboratory experiments elsewhere, where the rapid growth of fragments has been suggested to be a major contributing factor in the 'bloom' of *C. taxifolia* at new locations (Smith and Walters, 1999). Growth and survivorship of *C. taxifolia* fragments varied with salinity and temperature. In order to consider the relevance of these laboratory experiments in relation to existing conditions in Lake Conjola, long-term information of temperatures and salinities were examined using water quality data collected for Lake Conjola by Shoalhaven City Council (SCC, 2003).

Under laboratory conditions, the growth of fragments was negligible at 15°C but was high (>50 mm week^{-1}) at 20 and 25°C. Other laboratory experiments have similarly shown that other strains of *C. taxifolia* had little growth at temperatures between 10 and 15°C (Gayol et al., 1995;

Table 2 Results of ANOVAs to test for differences in mortality of *C. taxifolia* fragments (% bleached) grown in laboratory conditions at four separate temperatures

Temperature		15°C		20°C		25°C		30°C	
Source	df	MS	F	MS	F	MS	F	MS	F
Salinity (S)	4	3.78	41.1***	3.02	147.6***	3.26	59.4***	2.92	21.7***
Fragment (F)	2	0.01	0.10^{ns}	0.01	0.3^{ns}	0.08	1.5^{ns}	0.24	1.8^{ns}
S × F	8	0.04	0.42^{ns}	0.02	1.2^{ns}	0.04	0.8^{ns}	0.11	0.8^{ns}
Residual	30	0.09		0.02		0.05		0.13	
Cochran's		ns		ns		ns		*	
Tukey's HSD		17.5 > 20 > 22.5 = 25 = 30 ppt		17.5 > 20 = 22.5 > 25 = 30 ppt		17.5 > 20 > 22.5 = 25 = 30 ppt		17.5 = 20 = 22.5 > 25 = 30 ppt	

^{ns} not significant; *<0.05; **<0.01; ***<0.001

Tukey's HSD test has been used to indicate significantly different means. Notes: (1) data for salinity of 15 ppt was excluded from the analyses as there was 100% mortality; (2) growth data has been arcsin(x/100) transformed

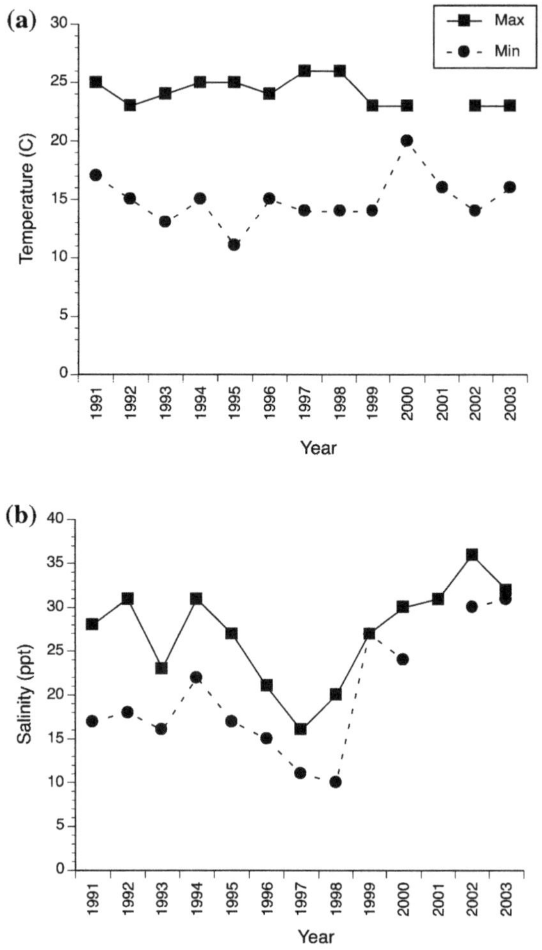

Fig. 2 Long-term maximum and minimum temperature (**a**) and salinity (**b**) records for Lake Conjola, NSW, Australia. Data have been made available from the Shoalhaven City Council (SCC, 2003). These data are for Roberts Point, located in the central basin of Lake Conjola (SCC, 2003, Site 43), for the period between May 1989 and June 2003

Komatsu et al., 1997; Chrisholm et al., 2000). The natural water temperatures of Lake Conjola are seasonal, with summer temperatures of approximately 23–25°C and winter temperature of between 12 and 16°C (Fig. 2a; SCC, 2003). Therefore, the results of the laboratory experiments in this study indicate that the *in situ* growth of *C. taxifolia* in Lake Conjola would be expected to be negligible at the recorded range of winter water temperatures, but high at the range of summer water temperatures. Field observations confirm this seasonality in the growth of *C. taxifolia* both locally (E. West, pers obs.) and in the Mediterranean situation (Ceccherelli and Cinelli, 1999a, b). These initial laboratory experiments proved very useful in determining the effect of salinity change on growth and mortality of *C. taxifolia*, and suggest that a range of experiments investigating other environmental factors, such as temperature, light climate, nutrient additions and so on, would be beneficial in understanding the limitations of this pest species in the region.

These experiments used *C. taxifolia* from Lake Conjola, which has an entrance that is intermittently open and closed to the Pacific Ocean (West et al., 1985; Pollard, 1994). Historically, Lake Conjola has undergone periods of entrance opening and closing, including closures that have lasted several years. Long term data for water quality reflects these periods of opening and closing in that salinity is close to marine (35 ppt) when open and falls during periods of closure (Fig. 2). In 2001, Lake Conjola was artificially opened to the sea and, despite drought conditions, has remained permanently opened since that time. For the 12 years of data prior to 2001, salinity at most sites was 30 ppt or below. During a long period of lake closure from 1995 to 1998, salinity fell to below 20 ppt and for about 2 years during this period was below 17 ppt (Fig. 2).

Since the lake has been permanently open, salinity has remained above 30 ppt at most sites and the lake is mostly marine. The laboratory experiments demonstrated that *C. taxifolia* fragments taken from the lake have a low growth rate and higher mortality at salinities below 22.5 ppt. Salinities in the range of 15–17.5 ppt resulted in almost total mortality of all fragments, at temperatures between 15 and 30°C, under growth cabinet conditions. This would suggest that *C. taxifolia*, which is primarily a marine alga, would be severely impacted by long periods of low salinity, similar to those that have occurred naturally in previous years prior to entrance manipulation (Fig. 2). Overall, the combination of laboratory experiments on the growth of *C. taxifolia* fragments and the long-term data for water quality, indicate that manipulation of the entrance may be an important factor in the long-term survival of this invasive alga at Lake Conjola.

Artificial opening of the entrance has maintained salinities that are close to seawater, which appear optimum for the growth of this predominantly marine alga. While it is uncertain that this has directly caused the establishment of *C. taxifolia*, there is little doubt that the continuation of these marine conditions assists in its survival and growth. One option for management that could be considered is the return of Lake Conjola to its natural opening regimes, which may in the future include long periods of closure and lowered salinity. On the evidence of these laboratory experiments and existing literature, entrance closure, if associated with lowered salinity, would have a negative impact on growth and may lead to mortality of large areas of the invasive alga. Prior to taking such a potentially controversial and unpopular decision, further laboratory and field-work may be required. These findings should also be considered when deciding on artificially opening any estuaries near infested sites in the region, because permanently opening an estuary could create nearly marine conditions, which appear to be ideal for establishment and further invasion of *C. taxifolia*. It is somewhat ironic that a managerial strategy used to enhance the health of the lake is potentially sustaining one of its greatest threats.

Acknowledgements We thank Andy Davis, Jeff Wright and Peter Barnes for their assistance, support and editorial comments. We also acknowledge Shoalhaven City Council for the use of their data and NSW Fisheries for their financial support.

References

Ashton, P. J. & D. S. Mitchell, 1989. Aquatic plants: patterns and modes of invasion, attributes of invading species and assessment of control programmes. In Drake J. A., H. A. Mooney, F. di Castri, R. H. Groves, F. J. Kruger, M. Rejmanek & M. Williamson (eds) Biological invasions: a global perspective. John Wiley and Sons, UK, 111–147.

Belsher, T. & A. Meinesz, 1995. Deep-water dispersal of the tropical alga *Caulerpa taxifolia* introduced into the Mediterranean. Aquatic Botany 51: 163–169.

Boudouresque, C. F, A. Meinesz, M. A. Ribera & E. Ballesteros, 1995. Spread of the green alga *Caulerpa taxifolia* (Caulerpales, Chlorophyta) in the Mediterranean: possible consequences of a major ecological event. Scientia Marina 59: 21–29.

Boudouresque, C. F. & M. Verlaque, 2002. Biological pollution in the Mediterranean Sea: invasive versus introduced macrophytes. Marine Pollution Bulletin 44: 32–38.

Ceccherelli, G. & F. Cinelli, 1999a. The role of vegetative fragmentation in the dispersal of the invasive alga *Caulerpa taxifolia* size in the Mediterranean. Journal of Experimental Marine Biology and Ecology 182: 299–303.

Ceccherelli, G. & F. Cinelli, 1999b. Effects of *Posidonia oceanica* canopy on *Caulerpa taxifolia* size in a north-western Mediterranean bay. Journal of Experimental Marine Biology Ecology 240: 19–36.

Ceccherelli, G., L. Piazzi & D. Balata, 2002 Spread of introduced *Caulerpa* species in macroalgal habitats. Journal of Experimental Marine Biology Ecology 280: 1–11.

Chrisholm, J. R. M., M. Marchioretti & J. M. Jaubert, 2000. Effects of low water temperature on metabolism and growth of a subtropical strain of *Caulerpa taxifolia* (Chlorophyta). Marine Ecology Progress Series 201: 189–198.

Gayol, P., C. Falconetti, J. R. M. Chisholm & J. M. Jaubert, 1995. Metabolic response of low water temperature acclimated *Caulerpa taxifolia* (Chlorophyta) to rapidly elevated temperature. Botanica Marina 38: 61–67.

Hurlbert, S. H., 1984. Pseudoreplication and the design of ecological field experiments. Ecological Monographs 54: 187–211.

Komatsu, T., A. Meinesz & D. Buckles, 1997. Temperature and light responses of the alga *Caulerpa taxifolia* introduced into the Mediterranean Sea. Marine Ecology Progress Series 146: 145–153.

Jousson, O., J. Pawlowski, L. Zaninetti, F. W. Zechman, F. Dini, G. Di Guiseppe, R. Woodfield, A. Millar & A. Meinesz, 2000. Invasive alga reaches California. Nature 408: 157.

Meinesz, A., 2002. Summary of Mediterranean invasion and management. Proceedings of the International *Caulerpa taxifolia* Conference 2002. San Diego, California (from the Sea Grant Non-Indigenous Site, www.sgnis.org).

Meinesz, A. & C. F. Bouderesque, 1997. On the origin of *Caulerpa taxifolia* in the Mediterranean Sea. Vie-Life Science 319: 603–613.

Meinesz, A., J. de Vaugelas, B. Hesse & X. Mari, 1993. Spread of the introduced tropical green alga *Caulerpa taxifolia* in northern Mediterranean waters. Journal of Applied Phycology 5: 141–147.

Phillips, J. A. & I. R. Price, 2002. How different is Mediterranean *Caulerpa taxifolia* (Caulerpales: Chlorophyta) to other populations of the species? Marine Ecology Progress Series 238: 61–71.

Pollard D.A., 1994. Opening regimes and salinity characteristics of intermittently opening and permanently open coastal lagoons on the south coast of New South Wales. Wetlands (Australia) 13: 16–35.

Ribera, M. A., E. Ballesteros, C. F. Boudouresque, A. Gomez & V. Gravez, 1996. Proceedings of Second International Workshop on *Caulerpa taxifolia*, Barce-

lona, 15–17 December 1994. University of Barcelona, Spain.
SCC, 2003. State of the Environment Report. Shoalhaven City Council, Nowra, Australia.
Schaffelke, B., N. Murphy & S. Uthicke, 2002. Using genetic techniques to investigate the sources of the invasive alga *Caulerpa taxifolia* in three new location in Australia. Marine Pollution Bulletin 44: 204–210.
Silva, P. C., 2002. Overview of the genus *Caulerpa.* Proceedings of the International *Caulerpa taxifolia* Conference, 2002. San Diego, California (from the Sea Grant Non-Indigenous Site, www.sgnis.org).
Smith, C. M. & L. J. Walters, 1999. Fragmentation as a strategy for *Caulerpa* species: fates of fragments and implications for management of an invasive weed. Marine Ecology 20: 307–319.
Vroom P. S. & C. M. Smith, 2001. The challenge of siphonous green algae. American Scientist 89: 525–531.
West, R. J., C. A. Thorogood, T. R. Walford & R. J. Williams, 1985. An estuarine inventory for New South Wales, Australia. Fisheries Bulletin 2. NSW Dept. Agriculture and Fisheries.
Zar, J. H., 1984. Biostatistical Analysis. Second Edition. Prentice-Hall International Inc, New Jersey USA.
Zuljevic, A. & B. Antolic, 2000. Synchronous release of male gametes of *Caulerpa taxifolia* (Caulerpales, Chlorophyta) in the Mediterranean Sea. Phycology 39: 157–159.

Hydrobiologia (2007) 577:95–106
DOI 10.1007/s10750-006-0420-9

LAGOONS AND COASTAL WETLANDS

Studies on the zooplankton community of a shallow lagoon of the Southern Baltic Sea: long-term trends, seasonal changes, and relations with physical and chemical parameters

Martin Feike · Reinhard Heerkloss · Thorsten Rieling · Hendrik Schubert

Abstract The Darß-Zingst Lagoon, a coastal inlet of the southern Baltic Sea, was subject to extended monitoring. The biomass data of zooplankton from 1969 to 2001 were used to analyze long-term trends and to correlate zooplankton biomass with abiotic factors. The dominant species in the lagoon were the calanoid copepods *Eurytemora affinis* and *Acartia tonsa*, and the rotifer *Keratella cochlearis* f. *tecta*. In the long-term trend, two pronounced changes in zooplankton biomass and species composition were observed. They are discussed in connection with a shift in dominance from macrophytes to phytoplankton and the invasion of a polychaet species into the lagoon. Significant relations between zooplankton data and abiotic parameters were found. While temperature, precipitation and NAO winter index correlated positively with copepods and negatively with rotifers, the relationships were inversely for pH-value and duration of ice cover.

Guest editors: P. Viaroli, P. Lasserre & P. Campostrini
Lagoons and Coastal Wetlands in the Global Change Context: Impacts and Management Issues

M. Feike (✉) · R. Heerkloss · T. Rieling · H. Schubert
Institute of Biosciences/Aquatic Ecology, University of Rostock, Albert-Einstein-Str. 3, D-18051 Rostock, Germany
e-mail: martin.feike@uni-rostock.de

Keywords Zooplankton · Coastal lagoon · Temperature · pH · Long-term time series · Seasonal changes

Introduction

The German coast of the Baltic Sea is characterized by several tideless lagoons (Boddens, Haffs) separated from the open sea by islands and peninsulas. These landlocked brackish waters are of high economic importance and used as sources of water for human use, i.e. fishery, recreation, agriculture, and waste water disposal. High external nutrient loading by the tributaries and diffuse sources from intensive agriculture caused heavy eutrophication of these shallow ecosystems. Within a complex program for maintaining the coastal water resources, the University of Rostock has been entrusted with the investigation of the basic hydrographical and biological characteristics of these landlocked coastal waters. The waters south of the Darß-Zingst peninsula were selected as investigation site. Since its implementation in 1969, a continuous monitoring program for hydrographical and biological parameters was performed (Schiewer, 1990).

The Darß-Zingst Lagoon (Fig. 1) has a length of 55 km, an area of 196.8 km^2 and a catchment area of 1600 km^2. The freshwater input supplied by the rivers Recknitz and Barthe is 61%

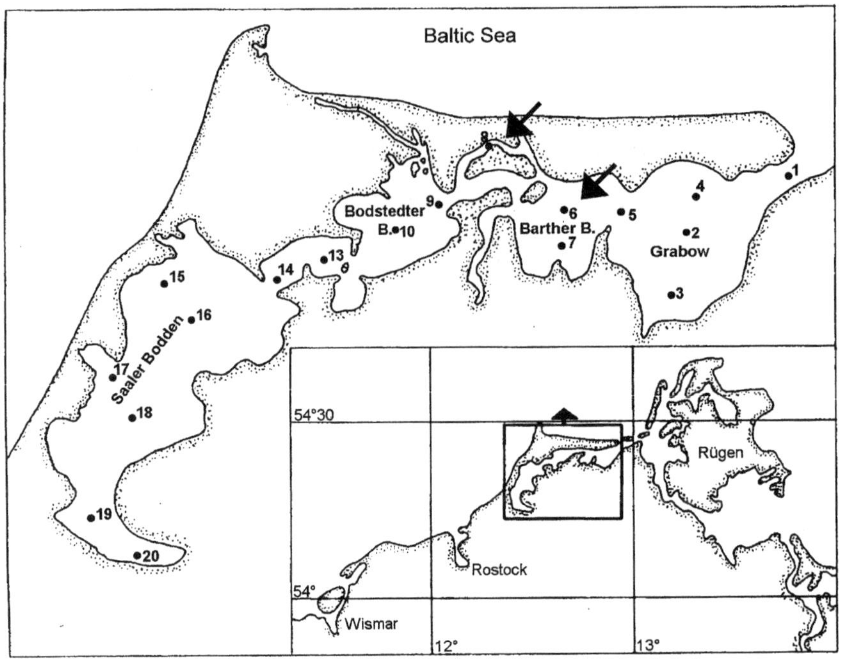

Fig. 1 Map of the Darß-Zingst lagoon. Numbers indicate the stations of monitoring. The arrows show to Station 8 (Zingster Strom) and station 6 (Barther Bodden), which are the sampling points of this investigation

(Schlungbaum et al., 1994a). According to Brosin (1965) the annual water balance is characterized as follows: Total volume 342×10^6 m^3, river run off 322×10^6 m^3, precipitation 115×10^6 m^3, evaporation 112×10^6 m^3, input from the sea 1021×10^6 m^3, output to the sea 1346×10^6 m^3. The considerable input from the sea during flood episodes is important for the trophic state. By loading with nutrients while mixing with the lagoon's water, the water from the Baltic Sea removes nutrients from the lagoon when flowing back. By diluting the lagoons water in this way, it improves its quality (Schnese, 1975). The exchange of water takes place slowly because the basins are connected by narrow channels. By the mixing of river water entering the Saaler and the Barther Bodden with the water coming from the Baltic Sea, the chain of basins does not act as a uniform water body, but the yearly retention times differ between the basins: Grabow 0.03 year^{-1}, Barther Bodden 0.05 year^{-1}, Bodstedter Bodden 0.03 year^{-1}, Saaler Bodden 0.14 year^{-1} (Schlungbaum et al., 1994b). Each basin has its own typical salinity regime, ranging from almost freshwater in the west to nearly 14 psu in the east. The increase in salinity is accompanied by decreases in nutrient concentrations, turbidity, algal biomass and primary production. According to data compiled by Wasmund (1990) for the period 1969–1980, from east to west the mean annual phytoplankton biomass increased from 3.6 to 27.9 mm^3 l^{-1} and the annual net primary production of phytoplankton increased from 109 to 611 g C m^{-2} a^{-1}, indicating a trophic status from slightly eutrophic to polytrophic. Wasmund & Schiewer (1994) reported values for the annual primary production of the Barther Bodden for a sequence of 14 years from 1971 to 1986. The arithmetic mean was 245.8 g C m^{-2} a^{-1} with a standard deviation of 72.8 g C m^{-2} a^{-1} and a median value of 215 g C m^{-2} a^{-1}. Due to the high productivity, the pH rose up to values of more than 9.5 in early summer with an inhibiting effect on the zooplankton development (Heerkloss & Schnese, 1999). Although both phytoplankton and zooplankton increased in biomass and production from east to west, a constant phytoplankton to zooplankton biomass ratio of 10:1 was observed (Schnese, 1973). The present paper analyses the long-term time series (1969–2001) of the zooplankton community structure and biomass. Special attention is paid at analyzing long-term changes in biomass, species composition and seasonal pattern, and at evaluating the relationships among different zooplankton groups and abiotic factors.

Methods

Sampling

Samples were taken at stations located in the central part of the lagoon (Fig. 1). The sampling method during the first 12 years of observation differed slightly from later times with respect to two points:

A Ruttner sampler was used from 1969 to 1980. From a depth of 0.5 m, 1 l was poured into a bottle and fixed with neutralized formaldehyde to a final concentration of 3%. A Fridinger sampler (Hydrobios, Kiel) was used since 1981 to avoid losses of fast-moving large animals due to the turbulence, which is induced by the horizontal folds of the Ruttner sampler. Although no significant difference in catching efficiency was found for the species occurring in the Darß-Zingst Lagoon, an underestimation of the larger copepod items cannot be fully excluded. However, the decline of copepods reported in this paper since the early 1980s is not considered to be an artefact produced by changing the type of sampler. In case of a methodological flaw, the result would be an increase in the biomass of the copepods, not a decrease as it was observed.

An integrated mode of sampling was introduced in 1981. Around 5×5 l was taken from both a depth of 0.5 m and from above the bottom. The samplers were emptied into a 50 l vessel. The content of 1 l was poured into a bottle after mixing the content of this vessel (full sample for small animals). Another 3 l were sieved through gauze with a mesh size of 50 μm. The animals were then washed into a small volume of 0.45 μm filtered lagoon water (net sample for large animals). Neutralized formaldehyde was added to both bottles to a final concentration of 3%. To make the results of the non-integrated sampling from the period before 1981 more representative and comparable to the integrated sampling, the mean biomass values from two locations (stations 8 and 6) were used for the period 1969–1980. (Fig. 1). The data since 1981 are all from station 8. The inclusion of the sampling station 6 could have had a diminishing influence on the biomass values used for our investigation. Schnese (1973) found that zooplankton biomass decreased from the western to the eastern parts of the lagoon. However, if there is an effect, it is rather slight in comparison to changes, which were observed in the long-term trend. A comparison between both locations from April to September in 1981–1985 resulted in average biomass values of 1.27 and 1.11 $mm^3\ l^{-1}$ at stations 6 and 8, respectively. The difference was not significant.

Determination of biomass

Standard methods were used for the microscopic counting and biomass calculation. Samples were counted under an inverted microscope or in Kolkwitz chambers. Geometric equivalents of the body volume of individuals were used for the calculation of the biomass from abundance data. A list of them is given by Heerkloss et al. (1991a). The seasonal variation in the body size of the copepods was taken into account. On an average, 2–3 samples were taken per month. In total approximately 1050, samples were investigated.

Data treatment and statistical analysis

To evaluate the joint influence of several parameters on zooplankton, a multivariate analysis was performed by Canonical Correspondence Analysis using the program CANOCO 4.51. The *P*-value was obtained by a Monte Carlo permutation test (5000 permutations), carried out for all canonical axes.

Results

Long-term changes in biomass and species composition

The biomass data were assigned to four taxonomic groups: the calanoid copepods, the rotifers, the larvae of the invading polychaete species *Marenzelleria neglecta* Sikorski and Bick and the group of the remaining species. In an earlier paper we designated *Marenzelleria neglecta* (Heerkloss & Schnese, 1999) as *Marenzelleria viridis* (Verrill). However, after a revision of the genus *Marenzelleria* by Sikorski & Bick (2004) it is now described as a new species. The long-term developments of the different groups are shown in Figs. 2, 3, 5, 6. The group of calanoid copepods was composed of only two species, *Eurytemora affinis* (Poppe), a typical species of estuaries in the northern hemisphere (Gasparini et al., 1999) and *Acartia tonsa* Dana, a North American species which invaded the Baltic Sea as a result of ballast water release (Naumenko, 2000). A sharp decline in biomass in the beginning of the 1980s was observed (Fig. 2). It is not clear whether the biomass of both species was included in this decrease because they were not differentiated in the microscopic countings of samples until 1980. It is, however, clear that *E. affinis* was involved in the decline because the spring peak fell clearly.

No change of the rotifer biomass was observed in that period, but a decline took place at the beginning of the 1990s (Fig. 3). The species diversity decreased in this time as well (Fig. 4), and *Keratella cochlearis* f. *tecta* (Gosse) became the most dominant species, constituting up to more than 95% of the total biomass of the rotifers.

Changes were observed also for the two other groups. As a result, of the invasion of the polychaete *M. neglecta*, meroplanktic larvae of this species developed high biomasses in late autumn and winter (Fig. 5). This Ponto-Caspian polychaet species invaded the Darß-Zingst Estuary in 1985 (Bick & Burckhardt, 1989). The larvae disappeared almost completely from the plankton after 1999, indicating a strong decrease of the polychaete.

The remaining zooplankton group consisted of phyllopods, cyclopoid and harpacticoid copepods, ostracods and meroplanktic larvae, but excluding the larvae of *M. neglecta*, the importance of this group being low during most of the time (Fig. 6). This group developed considerable biomass peaks in early summer only from 1969 to 1983. During those years, fresh water crustaceans of the group of the phyllopods were an important part of the zooplankton; especially *Chydorus sphaericus*

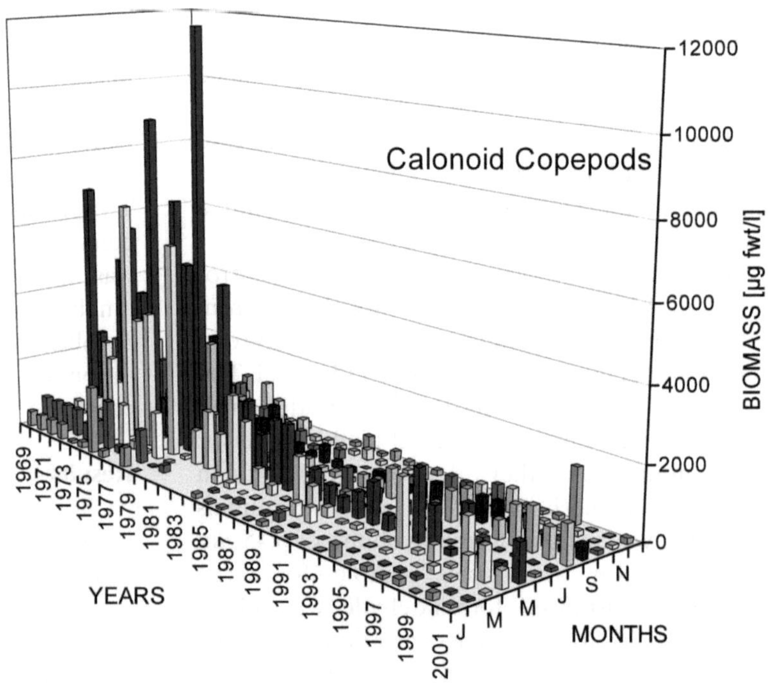

Fig. 2 Development of the biomass of the group of Calanoid copepods, consisting of the two species *E. affinis* and *A. tonsa*. Fwt = fresh weight

Fig. 3 Development of the biomass of the group of the Rotifers

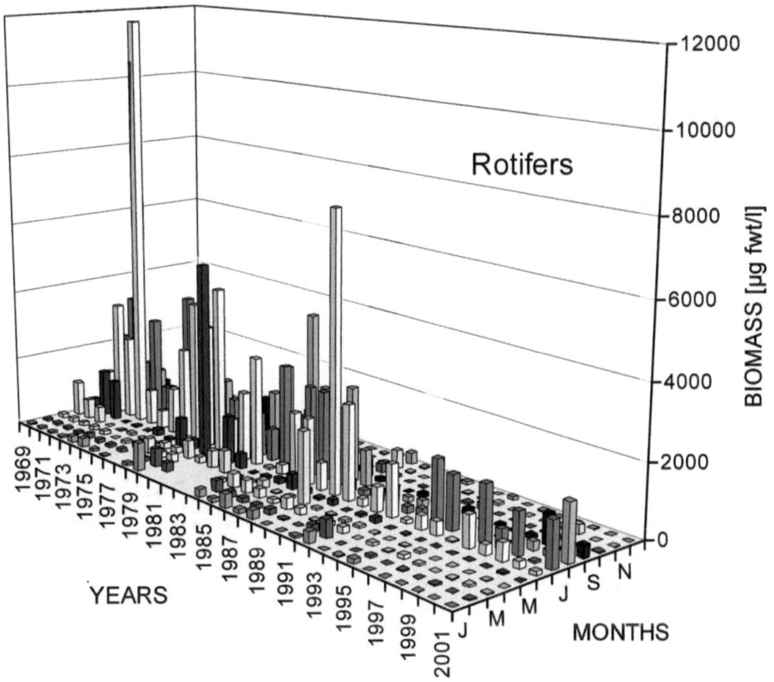

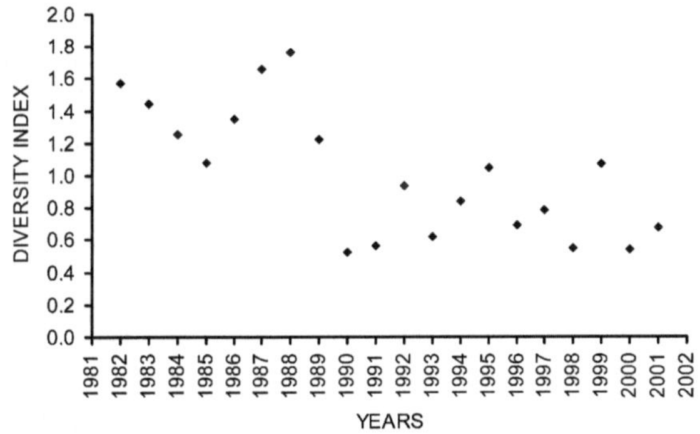

Fig. 4 Species diversity of the rotifer community at the station 8 (Zingster Strom). The Shannon-Weaver Index was calculated using the dual logarithm of the share of different species in total abundance

(O.F.M.), *Bosmina longirostris* (O.F.M.) and *Pleopsis polyphemoides* (Leuckart) were among the most dominant species. The phyllopods disappeared completely after that period. In 1970s, very high biomass values of phyllopods were found at low salinity. Salinity was also low in 1986–1988 and 1999 (see Table 1 for annual means of salinity), but no phyllopods were observed in those years. The disappearance of phyllopods can thus not be explained by a shift in the salinity regime of the estuary.

Because a seasonal pattern of biomass change is not clearly evident in Figs. 2, 3, 6, mean monthly values are shown in Figs. 7, 8. *Marenzelleria neglecta* is not included in these diagrams because its larvae were abundant only for a restricted period. The copepods showed clearly two peaks, the first in April and the second in late summer (Fig. 7). The spring peak consisted mainly of *E. affinis* whereas the second peak was composed also to a considerable degree by *A. tonsa* (Fig. 8). The rotifer biomass was highest in midsummer, filling up the gap between both copepod peaks (Fig. 7). The most abundant species among the rotifers was *K. cochlearis* f. *tecta* (Fig. 8). The principal pattern of seasonal change of biomass of the calanoid copepods and rotifers was stable during the whole period of observation (Figs. 2, 3).

Correlations with climatic, physical and chemical parameters

Besides long-term trends, an additional inter-annual variability was observed which may be depend on either the climatic features of the different years or on biotic interactions within the ecosystem. The Canonical Correspondence

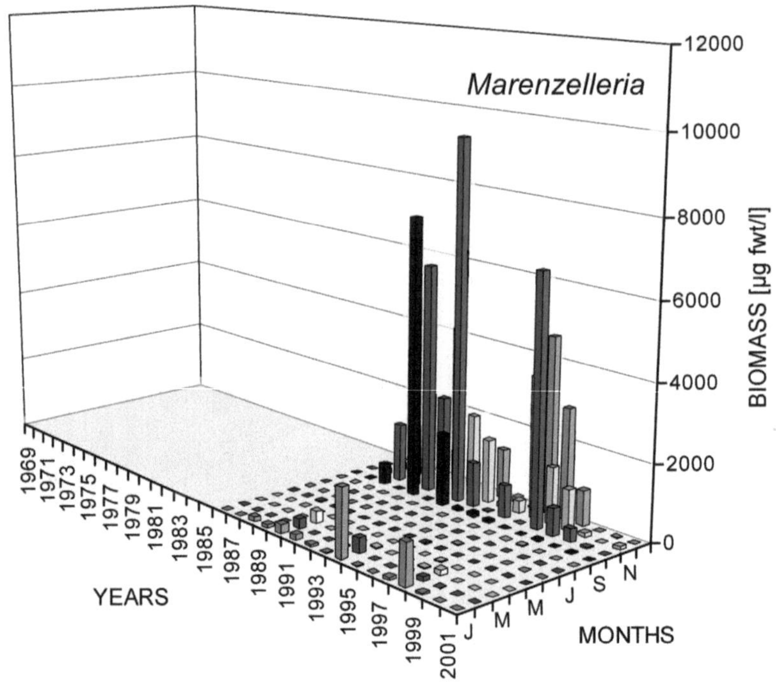

Fig. 5 Development of the biomass of the meroplanktonic larvae of the Ponto-Caspian polychaet species *Marenzelleria neglecta* which invaded the Darß-Zingst Estuary in 1985

Analysis (CCA) showed that the annual medians of monthly zooplankton biomasses were correlated with the abiotic data from 1986 to 2001 (Table 1, Fig. 9). A *P*-value of 0.016 was obtained by a Monte Carlo permutation test (5000 permutations), carried out for all canonical axes which indicates a high statistical significance of the relationships between the zooplankton and environmental variables. The first two axes explained 93.7% of the variance of the zooplankton data, of

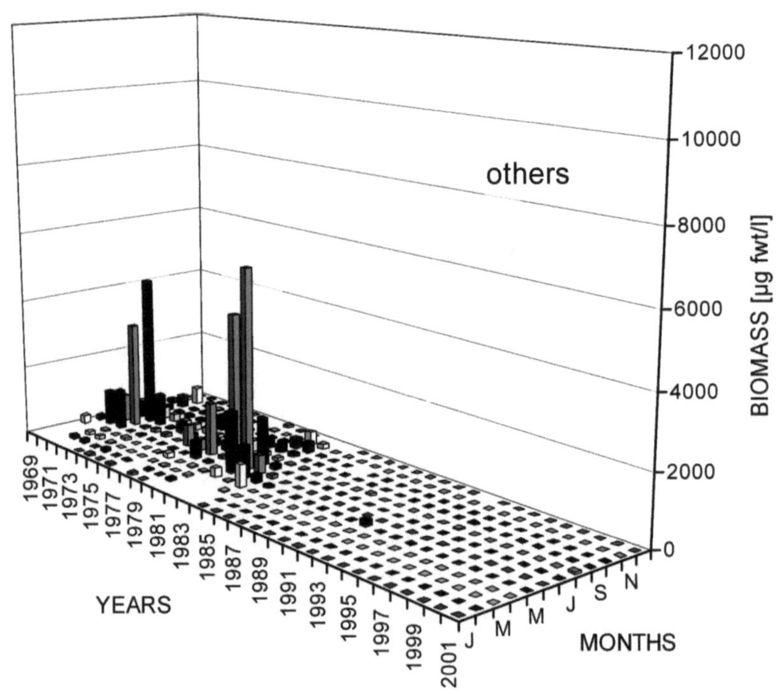

Fig. 6 Development of the biomass of the group of the remaining zooplankton species. This group consists of phyllopods, cyclopoid and harpacticoid copepods, ostracods, as well as mero-planktonic larvae except *Marenzelleria neglecta*

Table 1 Values of the abiotic parameters which are used for the multivariate statistical analysis in Fig. 9

Year	Temperature [°C]	NAO-index (winter)	Ice covering [d]	Salinity [PSU]	Precipitation [mm year^{-1}]	pH	NO_3^- [μM]	NO_2^- [μM]	NH_4^+ [μM]	PO_4^{3-} [μM]
1986	9.0	0.50	81	4.5	561.7	8.56	43.18	1.06	5.56	0.50
1987	8.5	–0.75	97	4.3	405.0	8.57	32.82	0.95	7.71	0.62
1988	9.7	0.72	6	4.4	525.5	8.72	91.08	1.24	7.40	0.53
1989	10.5	5.08	0	7.5	520.1	8.72	15.73	0.93	3.21	0.62
1990	10.8	3.96	2	7.5	618.6	8.97	19.00	3.12	5.19	0.66
1991	9.1	1.03	44	5.5	518.7	8.82	9.26	2.16	5.46	0.66
1992	9.9	3.28	10	7.2	520.0	8.61	6.49	0.64	4.37	0.71
1993	8.3	2.67	27	7.5	613.7	8.54	18.84	1.04	13.62	0.55
1994	9.1	3.03	37	5.0	646.9	8.69	49.13	1.60	11.43	0.58
1995	9.3	3.96	21	5.7	552.1	8.46	24.13	1.20	6.66	0.50
1996	8.1	–3.78	131	5.9	478.2	8.44	5.85	0.58	12.94	0.62
1997	9.5	–0.20	59	6.8	547.7	8.44	6.88	0.78	5.87	0.71
1998	9.1	0.72	18	5.8	790.2	8.65	28.40	1.08	5.29	0.61
1999	9.7	1.70	42	4.6	640.1	8.59	28.43	0.88	4.40	0.63
2000	10.2	2.80	8	5.8	581.2	8.48	18.35	0.81	4.02	0.53
2001	9.8	–1.89	12	5.1	700.4	8.47	13.03	0.62	3.87	0.38

Data are from Hurrell (1995) for the NAO winter index, Schumann et al. (2006) for salinity, Schlungbaum et al. (1994a), Baudler (2002, 2006) for temperature, precipitation, ice covering and nutrients. Except NAO index, all data are from station 8 (Zingster Strom)

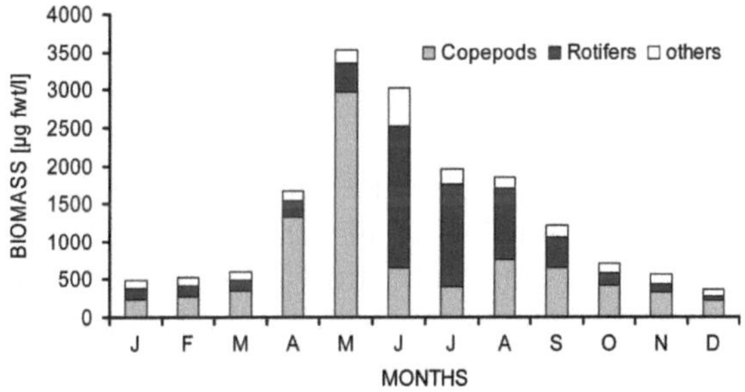

Fig. 7 Seasonal succession of zooplankton biomass, except *Marenzelleria neglecta*. Mean values for 1969–2001

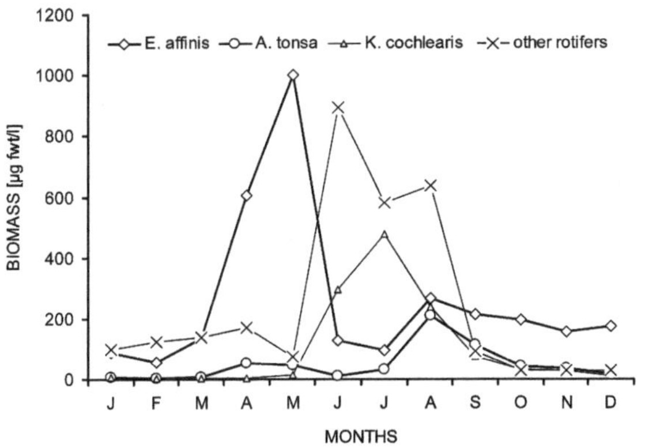

Fig. 8 Seasonal succession of biomass of some species of copepods and rotifers. Mean values for 1983–2001

which 99.7% of this variance was accounted by the first axis and only 0.3% by second axis. The most important factors determining the distribution in the diagram of the first two canonical axes were temperature, precipitation, pH-value and, to a lesser extent, the duration of ice cover and the NAO winter index. High values of the latter parameter were indicators of years, which started with a mild and damp winter weather. The optimum of rotifers corresponded well with high pH-values and a long duration of ice cover, whilst copepod dominance corresponded to high values of temperature, precipitation and NAO index.

Discussion

The Darß-Zingst Estuary consists of very shallow, unsheltered and semi-enclosed basins, which share some features with shallow lakes. The water column in such lakes is always well-mixed, and the turbulence induced by wind provides for a high content of aged and re-suspended detritus in the water column. The phenomenon of alternative stable states, in which the systems is dominated either by submerse macrophytes or by

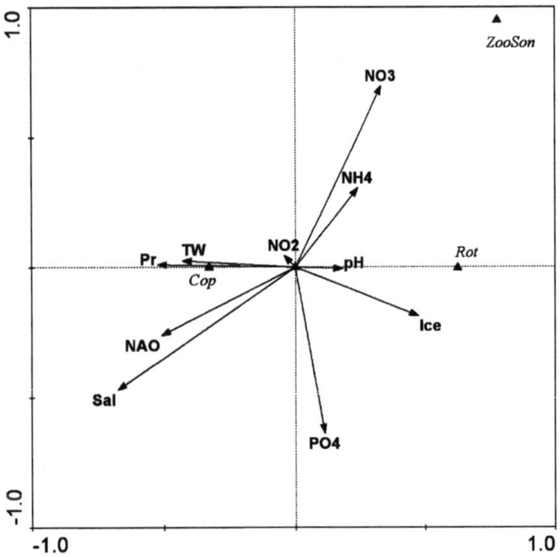

Fig. 9 Result of a CCA for the period 1986–2001. Annual median values of zooplankton. Cop = group of calanoid copepods, Rot = group of rotifers, Other = group of remaining zooplankton, Pr = precipitation, Sal = salinity, Ice = duration of ice cover in the winter before the year, T = temperature, PO4 = soluble reactive phosphorous, NH4 = ammonium, NO2 = nitrite, NO3 = nitrate, NAO = winter index of the North Atlantic Oscillation. The two axes of the CCA-plot explain 93.7% of the variance in the zooplankton data. 99.7% of this variance are represented by the first axis, 0.3% by the second axis

phytoplankton is also characteristic for shallow lakes. A switch from the macrophyte state to the phytoplankton state may occur in a rather short time when the macrophytes are reduced due to meteorological influences, e.g. a hard winter or strong river runoff (Scheffer et al., 1993). A similar mechanism might be responsible for some of the observed changes during the long-term monitoring reported in this study.

A massive decline of submersed macrophytes — mainly Characea — took place in the Darß-Zingst Lagoon in 1981 due to a rainy spring season with low water salinities, high nutrient loads, and high water turbidity (Schiewer, 1997). Under these conditions, the dominance shifted from macrophytes to phytoplankton, which resulted in increased turbidity and sediment mobility. In addition, in the central part of the Darß-Zingst Estuary an increase of the aged detritus took place, which accounted for 80% of the total organic seston (Georgi, 1985). In parallel, in summer C/N ratios of the seston between 10 and 15 were determined. The disappearance of the phyllopod species since the early 1980s could be a result of these changes in the suspended particulate matter because phyllopods are non-selective filter feeders. Another explanation would be an increase in the top-down control by vertebrate and non-vertebrate predators. However, this alternative can be excluded because the grazing pressure of planktivorous fish on zooplankton was found to be low in the Darß-Zingst Lagoon (Mehner & Heerkloss, 1994), and also because the data on the secondary production of the main non-vertebrate predator, the mysid *Neomysis integer* (Leach), indicate to a low top down control of zooplankton by this species (Jansen, 1983).

The copepod decrease was probably also depended on changes of seston composition. The herbivorous calanoid copepods are able to select food (Irigoien et al., 2000), but selectivity can be contrasted by seston composition, when the non living fraction exceeds the phytoplankton quota. *Eurytemora affinis* which is typical typical estuarine habitats suffers from high concentrations of dead material in the food supply (Tackx et al., 2004). The connection of low egg production rates of *E. affinis* with poor food quality was observed by Burdloff et al. (2000) in the Gironde estuary. A strong indication of a limitation of *E. affinis* by low food quality in the Darß-Zingst Lagoon was indicated by relationships between the C/N ratios of the seston and egg ratios. According to Georgi (1985), an increase of the C/N-value from 8.8 to 10.2 was observed from May to June. At the same time, the egg ratio decreased from 17 to 3 (Arndt, 1985). A strong drop of the feeding rate of *E. affinis* was also observed during this period (Heerkloss et al., 1990). Furthermore, since 1981 changes in the composition of phytoplankton community have to be taken into account as a cause of a decrease in food quality. Nanoplanktonic species which are selected by *E. affinis*, were partly replaced either by picoplankton cyanobacteria or by clumped colonies of the green algae *Tetrastrum triangulare* (Chod.) (Wasmund & Schiewer, 1994).

The decline in rotifer biomass took place a decade later, in the early 1990s, when nutrient loads from point and diffuse sources decreased considerably (Schlungbaum et al., 2001). This resulted in a significant recovery of mesotrophic conditions as indicated by phytoplankton (Schumann & Karsten, 2006). However, these changes took place gradually over a rather long period, whilst the decline of the rotifer biomass and species diversity occurred suddenly between 1988 and 1990. Therefore, it is unlike that the changes observed in rotifer communities were connected with water quality recovery. The hypothesis of an increased grazing mortality was supported neither by long-term data on ichthyofauna (Winkler, 2002) nor by data on carnivorous zooplankton. No data have been available since the late 1980s for the mysid *N. integer*, but the earlier results of Jansen (1983) evidenced a little top down effect of this predator. According to a rough estimation by Arndt (1985), the maximum feeding rate of *Neomysis* in 1979 and 1980 amounted to 50 mg fwt l^{-1} d^{-1} in June. The mean rotifer production in June in those years was 2206 mg fwt l^{-1} d^{-1}, i.e. only up to 2.3% of the production was consumed by *Neomysis*. To explain the strong decrease of rotifers after 1988 by *Neomysis* grazing, a rapid and very strong increase of this predator had to be assumed. A further change in the long-term data, which coincided with the rotifer decline, was the appearance of high abundances of meroplanktic larvae of the polychaete *M. neglecta* in 1988 (Fig. 5). This polychaete developed well in the Darß-Zingst Lagoon for 10 years after invasion. Its growth resulted in a 7-fold increase in the biomass of macro-zoobenthos (Arndt, 1994). This could have led to an intensification of bioturbation, which would result in a loss of the rotifers resting eggs by transport to deeper sediment layers. The worm is active down to a depth of 30 cm (A. Bick pers. comm.). Furthermore, a grazing on the resting eggs of rotifers could be a cause of their decline. Assuming size selective grazing on larger eggs, the shift in species composition to the very small sized *K. cochlearis* f. *tecta* may be also explained in this connection. It is thus the task of future investigations to find empirical support for this hypothesis. One consequential prediction is that changes in the opposite direction will take place when the population size of *M. neglecta* will decrease. Recently, in 2000 and 2001 very low biomass values of larvae were attained. The continuation of monitoring in the next years will show whether the polychaete will remain at a low level and, if so, whether changes in the rotifers will take place.

The seasonal succession of the zooplankton biomass showed a very stable pattern. *Acartia tonsa* always succeeded *E. affinis* during summer, a pattern which was also observed in other estuaries and lagoons (Bakker & De Pauw, 1975; Baretta & Malschaerd, 1988; Paffenhöfer & Stearns, 1988), namely in the Vistula Lagoon which is also a semi-enclosed, shallow and tideless coastal inlet of the Baltic Sea. The seasonal succession in biomass and species composition in the Vistula lagoon was similar to what observed in the Darß-Zingst Lagoon. However, no mass development of rotifers in summer was observed in the Vistula (Heerkloss et al., 1991b). Because the trophic state is lower in the Vistula Lagoon than in the Darß-Zingst Lagoon, the difference in the rotifer communities seemed to be related to eutrophication. Besides *K. cochlearis* f. *tecta*, the summer peak of rotifers in the Darß-Zingst Lagoon was due mainly to *Filinia longiseta* (Ehrenberg). Both species feed preferentially on bacteria and small detritus particles (Pourriot, 1977). Thus, the development of bacterio-detritivorous rotifer species under the conditions of advanced eutrophication in the Darß-Zingst Lagoon was an expression of intensified microbial activities (Schiewer, 1997). The high trophic state was also evidenced by the dominance of the form-type *K. cochlearis* f. *tecta* among all form-types of this species. The relative abundance of *K. cochlearis* f. *tecta* among all *K. cochlearis* forms has been shown to correlate positively with the trophic state (Karabin, 1985).

Correlations with abiotic factors

The available data from the long-term monitoring allow conclusions about the influence of climatic conditions on the biomass of the different zooplankton groups. The positive correlation of copepods with temperature, precipitation and NAO index indicated that the copepods develop better in warm years after a mild and dump winter. This will result in a promotion of fish recruitment because the numbers of juvenile fish correlate very well with copepod biomass (Winkler, 2002). The cause of the observed temperature effect on copepods may be related to the temperature requirements of *A. tonsa*, which is assumed to be thermophilic. Capuzzo (1979) determined a good tolerance to higher temperatures of this species in comparison of other marine copepods, and Kimmel & Roman (2004) observed a positive correlation of *A. tonsa* development with temperature in Chesapeake Bay. *Eurytemora affinis* is a eurythermic species (Tackx et al., 2004). For individuals from the Darß-Zingst Lagoon, Ring et al. (1985) estimated 15°C as the optimum for the feeding rate in summer, which is below the in situ temperature during the same time. *Eurytemora affinis* can thus not be considered as thermophilic. We performed an additional CCA analysis on the correlation of abiotic factors with several species, which resulted in no significant dependence of *E. affinis* on temperature. Therefore, the observed correlation in Fig. 9 has to be considered as a result of an increase of *A. tonsa* biomass with the annual mean temperature. The better development of copepods during warmer and of rotifers during colder years coincides with the investigation of Heerkloss et al. (2005) on temperature dependent changes of the Darß-Zingst Lagoon plankton. In looking for possible effects of global warming, that paper compared 5 years with a special warm climate with 5 cold years. It was shown that the annual mean biomass was significantly higher for copepods and lower for rotifers in warm years.

Ring et al. (1985 with laboratory experiments demonstrated that the feeding rate of *E. affinis* is inhibited at pH > 9.0, a pH level which often occurs in early summer in the Darß-Zingst Lagoon. The positive correlation of the pH-value with rotifers is in agreement with findings from laboratory experiments. A preference of alkaline conditions was found for *Brachionus calyciflorus* Pallas by Mitchell (1992) and *Brachionus plicatilis* (O.F. Müller) by Carius (1995). Both species occurred regularly in the Darß-Zingst Lagoon but seldom with high abundances. Our results might be a hint, that *K. cochlearis* f. *tecta* and *F. longiseta*, the rotifer species which are mainly responsible for the rotifer mass development in summer, have also a preference to high pH-values.

Conclusions

The long-term monitoring program of the Darß-Zingst Lagoon has revealed two periods of change in zooplankton. In the early 1980s, a change in the crustaceans was observed. The species richness decreased due to the disappearance of phyllopods and a drop of the average level of biomass of the copepods took place as a result of food quality degradation. A decade later, a drop of the rotifers biomass and a decrease in evenness by the mass development of *K. cochlearis* f. *tecta* took place. Because this event can be explained neither by bottom up nor by top down effects in the pelagic system, the considerable increase in macro-zoobenthos biomass by the invader *M. neglecta* might be responsible for the changes of rotifers. This finding needs further field observations and experimental work. Beside these two events, the observed biomass differences between years can be attributed to differences in the climatic conditions of the individual years. However, the main patterns of the seasonal biomass curves of copepods and rotifers seemed stable.

Acknowledgements The authors gratefully acknowledge Dr. Henning Baudler for the contribution of abiotic data as well as Friedel Fischer, Helga Albrecht, Sybille Müller and Gerda Hinrich for counting zooplankton samples and technical help. Many thank also to an unknown referee for helpful criticism and extensive suggestions on the improvement of the manuscript.

References

Arndt, E. A., 1994. Struktur und dynamik des makrozoobenthos in der Darß-Zingster boddenkette im laufe der letzten 25 jahre unter besonderer berücksichtigung des makrozoobenthos—entwicklung im saaler bodden von 1986–1990. Rostocker Meeresbiologische Beiträge 2: 93–120.

Arndt, H., 1985. Untersuchungen zur Populationsökologie der Zooplankter eines inneren Küstengewässers der Ostsee. Dissertation A, Universität Rostock, 170 pp.

Bakker, C. & N. De Pauw, 1975. Composition of plankton assemblges of identical salinity ranges in estuarine tidal, and stagnant environments. Netherlands Journal of Sea Research 2: 69–81.

Baretta, J. W. & J. F. P. Malscheart, 1988. Distribution and abundance of the zooplankton of the Ems estuary (North Sea). II. Zooplankton. Netherlands Journal of Sea Research 2: 69–81.

Baudler, H., 2002. Projekt KEI. Untersuchungen zur Kurzzeitvariabilität ausgewählter Eutrophierungs-Indikatoren im Zingster Strom. 01.09.2000–31.08.2002. Landesamt für Umwelt, Naturschutz und Geologie, 54 pp.

Baudler, H., 2006. Hydrographisches Monitoring der Laborstation Zingst. Rostocker Meeresbiologischer Beiträge 16: 35–45.

Bick, A. & R. Burckhardt, 1989. Erstnachweis von *Marenzelleria viridis* (Polychaeta, Spionidae) für den Ostseeraum, mit einem Bestimmungsschlüssel der Spioniden der Ostsee. Mitteilungen aus dem Zoologischen Museum in Berlin 65: 237–247.

Brosin, H. J., 1965. Hydrographie und wasserhaushalt der boddenkette des Darß und des zingst. Geophysikalische Veröffentlichungen der Karl-Marx-Universität Leipzig 18: 277–381.

Burdloff, D., S. Gasparini, B. Sautour, H. Etcheber & J. Castel, 2000. Is the copepod egg production in a highly turbid estuary (the Gironde, France) a function of the biochemical composition of seston? Aquatic Ecology 34: 165–175.

Capuzzo, J. M, 1979. The effect of temperature toxicity of chlorinated cooling waters to marine animals—a preliminary review. Marine Pollution Bulletin 10: 45–47.

Carius, H.-J., 1995. Autökologische Untersuchungen zur Reaktion Ausgewählter Zooplankter auf Abiotische und Biotische Umweltfaktoren Unter Besonderer Berücksichtigung der Wirkung von hohen pH-Werten und Mikroparasiten. Diplomarbeit, Universität Rostock, 75 pp.

Gasparini, S., J. Castel & X. Irigoien, 1999. Impact of suspended particulate matter on egg production of the estuarine copepod, *Eurytemora affinis*. ICES Journal of Marine Science 22: 195–205.

Georgi, F., 1985. Verteilung und beschaffenheit des sestons in inneren küstengewässern der DDR (Darß-Zingster Boddengewässer) sowie der westlichen und mittleren ostsee unter besonderer berücksichtigung der wasseraustauschprozesse zwischen beiden systemen. Beiträge zur Meereskunde 52: 35–48.

Heerkloss, R., U. Brenning & M. Ring, 1990. Secondary production of calanoids (Copepoda, Crustacea) in Brackish waters of the southern Baltic. Limnologica (Berlin) 20: 65–69.

Heerkloss, R., W. Schnese & B. Adamkiewicz-Chojnacka, 1991a. Eutrophication induced changes of zooplankton communities in coastal regions of the Baltic Sea. Internationale Revue der Gesamten Hydrobiologie 76: 397–404.

Heerkloss, R., W. Schnese & B. Adamkiewicz-Chojnacka, 1991b. Eutrophication induced changes of zooplankton communities in coastal regions of the Baltic Sea. Internationale Revue der Gesamten Hydrobiologie 76: 397–404.

Heerkloss, R. & W. Schnese, 1999. A long-term series of zooplankton monitoring of a shallow coastal water of the southern Baltic. Limnologica 29: 317–321.

Heerkloss, R., Th. Rieling & H. Schubert, 2005. Long-term studies of temperature dependent plankton community changes in an estuarine system of the southern baltic sea. icam dossiers nr. 3: Lagoons and coastal wetlands in the global change context: impact and management issues. Proceedings of the international conference, Venice, 26–28 April 2004: 159–164.

Hurrell, J. W., 1995. Decadal trends in the North Atlantic oscillation: regional temperatures and precipitation. Science 269: 676–679.

Irigoien, X., R. N. Head, R. P. Harris, D. Cummings, D. Harbour & B. Meyer-Harms, 2000. Feeding selectivity and egg production of *Calanus helgolandicus* in the English Channel. Limnology and Oceanography 45: 44–55.

Jansen, W., 1983. *Neomysis integer* (Leach) (Crustacea, Mysidacea) in der Darß-Zingster Boddenkette—Untersuchungen zur Populationsentwicklung und Leistungsfähigkeit in Abhängigkeit von Umweltfaktoren. Dissertation A, Universität Rostock, 134 pp.

Karabin, A., 1985. Pelagic zooplankton (Rotatoria, Crustacea) variation in the process of lake eutrophication. Ekologia Polska 33: 567–616.

Kimmel, D. G. & M. R. Roman, 2004. Long-term trends in mesozooplankton abundance in Chesapeake Bay, USA: influence of freshwater input. Marine Ecology Progress Series 267: 71–83.

Mehner, Th. & R. Heerkloss, 1994. Direct estimation of food consumption of juvenile fish in a shallow inlet of the southern Baltic. Internationale Revue der Gesamten Hydrobiologie 79: 295–304.

Mitchell, S. A., 1992. The effect of pH on *Brachionus calyciflorus* Pallas (Rotifera). Hydrobiologia 245: 87–93.

Naumenko, E. N., 2000. The dynamics of abundance of the introduced *Acartia tonsa* Dana in the Vistula Gulf of the Baltic Sea. In G. G. Matishov, V. V. Dnisov, A. D. Chinarina, V. S. Znzerov & E. G. Berestovky, (eds), Species Introducers in the European Seas in Russia. Kol'skij Nauchnyi Tsentr RAN, Apatity (Russia): 113–121.

Paffenhöfer, G.-A. & D. Stearns, 1988. Why is *Acartia tonsa* (Copepoda, Calanoida) restricted to near shore environments? Marine Ecology Progress Series 42: 33–38.

Pourriot, R., 1977. Food and feeding habits of rotifers. Archiv für Hydrobiologie, Beihefte Ergebnisse der Limnologie 8: 243–260.

Ring, M., R. Heerkloss & W. Schnese, 1985. Einfluss von temperatur, pH-Wert und nahrungsqualität unter laborbedingungen auf *Eurytemora affinis*. Wissenschaftliche Zeitschrift der Wilhelm-Pieck-Universität Rostock, Mathematisch-naturwissenschaftliche Reihe 34: 22–25.

Scheffer, M., S. H. Hosper, M.-L. Meijer, B. Moss & E. Jeppesen, 1993. Alternative equilibria in shallow lakes. Trends in Ecology and Evolution 8: 275–279.

Schlungbaum, G, H. Baudler & G. Nausch, 1994a. Die Darß-Zingster Boddenkette—ein typisches Flachwasserästuar an der südlichen Ostseeküste. Rostocker Meeresbiologische Beiträge 2: 5–26.

Schiewer, U., 1990. Werner Schnese and the development of coastal water ecology in Rostock, GDR. Internationale Revue der Gesamten Hydrobiologie 7: 1–13.

Schiewer, U., 1997. 30 years eutrophication in shallow brackish waters—lessons to be learned. Hydrobiologia 363: 73–79.

Schlungbaum, G., U. Schiewer & E. A. Arndt, 1994b. Beschaffenheitsentwickung und Klassifizierung der Darß-Zingster Boddengewässer mit ausgewählten Vergleichen zu anderen Bodden und Haffen. Rostocker Meeresbiologische Beiträge 2: 191–213.

Schlungbaum, G., H. Baudler, M. Krech & B. Kwiatkowski, 2001. Die Darß-Zingster bodden—eine studie. Schriftenreihe des Landesamtes für Umwelt, Naturschutz und Geologie. No 1, 209 pp.

Schnese, W., 1973. Relation between phytoplankton and zooplankton in brackish coastal water. Oikos, Supplement 15: 28–33.

Schnese, W., 1975. Estimation of plankton production in shallow inlets of the GDR Baltic coast line. Havsforskningsinstituts Skrift (Helsinki) 239: 131–136.

Schumann, R. & U. Karsten, 2006. Phytoplankton im Zingster Strom der Darß-Zingster Boddenkette–13 Jahre Remesotrophierung. Rostocker Meeresbiologische Beiträge 16: 47–59.

Schumann, R., H. Baudler, Ä. Glass, K. Dümcke & U. Karsten, 2006. Long-term observations on salinity dynamics in a tideless shallow lagoon of the Southern Baltic Sea coast. Journal of Marine Systems: 330–344.

Sikorski, A. V. & A. Bick, 2004. Revision of *Marenzelleria* Mesnil, 1896 (Spionidae, Polychaeta). Sarsia 89: 253–275.

Tackx, M. L. M., N. de Pauw, R. van Mieghem, F. zemar, A. Hannouti, S. van Damme, F. Fiers, N. Daro & P. Meire, 2004. Zooplankton in the Schelde estuary, Belgium and the Netherlands. Spatial and temporal patterns. Journal of Plankton Research 26: 133–141.

Wasmund, N., 1990. Charcteristics of phytoplankton in brackish waters of different trophic levels. Limnologica (Berlin) 20: 47–51.

Wasmund, N. & U. Schiewer, 1994. Überblick zur Ökologie und Produktionsbiologie des Phytoplanktons der Darß-Zingster Boddenkette. Rostocker Meeresbiologische Beiträge 2: 41–60.

Winkler, H. M., 2002. Effects of eutrophication on fish stocks in Baltic lagoons. In Schernewski G. & U. Schiewer (eds), Baltic Coastal Ecosystems. Springer-Verlag, Berlin, 65–74.

Hydrobiologia (2007) 577:107–125
DOI 10.1007/s10750-006-0421-8

LAGOONS AND COASTAL WETLANDS

Hydrographic, geomorphologic and fish assemblage relationships in coastal lagoons

Angel Pérez-Ruzafa · Mª Carmen Mompeán · Concepción Marcos

Abstract In this study, 40 Atlanto-Mediterranean coastal lagoons were analyzed in order to evaluate the extent to which their ecological characteristics depend on hydrographic, trophic or geomorphologic features. Fish species richness increases with lagoon volume and the openness parameter, which characterizes the potential influence of the sea on general lagoon hydrology and is related to the total transversal area of the inlets, which connect the lagoon to the sea. On the other hand, the number of species decreases exponentially with the phosphate concentration in water. The fishing yield increases with the chlorophyll *a* concentration in the water column and exponentially with shoreline development. With respect to the fish assemblage composition, geomorphologic features alone explain 22% of the variance in the canonical analyses and an additional 75% when including the hydrographic and trophic characteristics of the lagoon, the latter on its own explaining only 3% of the observed differences.

Guest editors: P. Viaroli, P. Lasserre & P. Campostrini
Lagoons and Coastal Wetlands in the Global Change Context: Impacts and Management Issues

A. Pérez-Ruzafa (✉) · Mª. C. Mompeán · C. Marcos
Department of Ecology and Hydrology, University of Murcia, 30100 Murcia, Spain
e-mail: angelpr@um.es

Keywords Coastal lagoon · Fish assemblages · Typification · Species richness · Fishing yield

Introduction

Coastal lagoons are dynamic ecosystems characterized by particular features such as shallowness, relative isolation from the open sea due to coastal barriers that maintain some communication channels or inlets, and the presence of boundaries with strong physical and ecological gradients (UNESCO, 1981). They are considered naturally stressed systems with frequent environmental disturbances and fluctuations (Barnes, 1980; Kjerfve, 1994; UNESCO, 1980, 1981).

Under the term lagoon comes a wide range of environments. Size can vary from a few hundred square meters to extensive areas of shallow coastal sea. Salinity ranges from nearly fresh to hypersaline waters (Barnes, 1980). The salt balance relies on several factors such as the exchange of water with the open sea, the input of continental waters from rivers, watercourses, groundwater, and the rainfall-evaporation balance. Salinity may also be variable within a lagoon, both spatially and temporally. From a hydrographic point of view, most of the variability between lagoons could be summarized in a set of quantitative parameters or indexes describing the lagoons' orientation (parallel or perpendicular to the shore) and structure, as well as spatial variability and a potential influence by the sea.

In order to explain how coastal lagoon ecosystems function, several proposals for the classification and typification of these environments have been made using distinct criteria such as salinity, substrate type, formation, isolation, size, morphology, etc. (Barnes, 1980; Guelorget & Perthuisot, 1983). Among these proposals, some of the more important ones used salinity as the main parameter (Petit, 1953; Por, 1972, 1980), with the Venice System (Anonymous, 1959) being one of the more widespread. In these models, coastal lagoons were considered a transitional system between continental and marine domains (Bianchi, 1988).

Kjerfve (1994) subdivided coastal lagoons into three geomorphic types, choked, restricted and leaky, as three points along a spectrum reflecting the exchange of water with the coastal sea. The rate and magnitude of oceanic exchange reflect both the dominant forcing functions and the time-scale of hydrological variability. Physical gradients within the lagoon environment derived from such exchange rates have also been related to biological gradients in species richness, abundance and productivity. In the early 1980's, Guelorget & Perthuisot (1983) and Guelorget et al. (1983) rejected salinity as an essential parameter for explaining the observed gradients in density, biomass, species richness or diversity and proposed that zonation patterns and species distribution inside the lagoons be determined by confinement, a parameter which represents the turnover time for marine waters and impoverishment in some oligo-elements of a marine origin. Later, Pérez-Ruzafa & Marcos (1992, 1993) suggested that, instead of the recycling of vitamins and oligo-elements, the main factor explaining the lagoon assemblage structure in a confinement gradient would be that of colonization rates by marine species. The species composition at each lagoon site would be the result of equilibria in the context of interspecific competition between marine and lagoon species, taking into account that low competition coefficients for alloctonous species may be compensated by high immigration rates from outer habitats.

In either case, most biological assemblage characteristics, including community structure and productivity, would be related to the geomorphologic characteristics of the lagoons. The biological assemblage structure is conditioned by environmental stress (Gamito et al., 2005). Species respond to these situations via life-history strategies and are usually species which show an increased reproductive effort by reproducing earlier, as well as having smaller and numerous offspring.

Due to their characteristics, coastal lagoons are usually among those marine habitats with the highest biological productivity (Allongi, 1998), functioning as nurseries and feeding grounds for opportunistic marine-estuarine fish and sustaining important fisheries (Clark, 1998; Yañez-Arancibia & Nugent, 1977). As a result, these environments are of great social concern and constitute one of the main priorities in the integrated management of coastal areas due to their susceptibility to human impacts and the intensification of competing uses.

Considering the need for managing and typifying these ecosystems and in the context of the European *Water Framework Directive-WFD*, the question arises as to what extent the ecological characteristics and variability of coastal lagoons can be explained by geomorphologic features? The relevance of this question lies in the fact that geomorphologic features such as depth, the influence of the sea, coastal development, lagoon size, as well as water temperatures and productivity are also clearly affected by the rise in sea level, global climate changes, and most human activity related with urban development, tourist facilities and the exploitation and management of these areas.

In this paper, we analyze the available information on fish assemblages and geomorphologic, hydrological and trophic features in a number of coastal lagoons in order to answer this question.

Materials and methods

Data from 40 lagoons in the Atlanto-Mediterranean area were compiled from bibliographic and cartographic sources (Fig. 1, Table 1). Thirty-six different lagoon variables were analyzed (Table 2

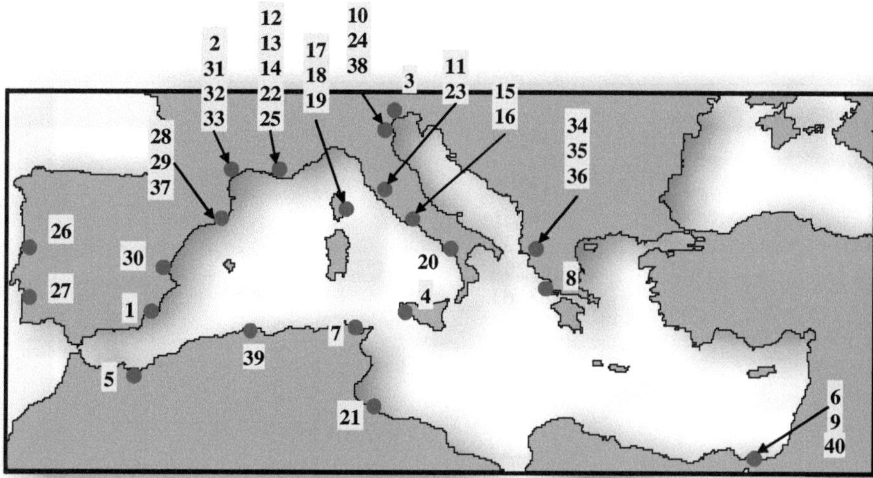

1	MM	**Mar Menor**	11	BUR	**Burano**	21	BIB	**Biban**	31	SL	**Salses-Leucate**
2	TH	**Thau**	12	PR	**Prevost(V-l-M)**	22	AR	**Arnel(V-I-M)**	32	VA	**Vaccarès**
3	VE	**Venecia**	13	MEJ	**Mejean(L)**	23	OR	**Orbetello**	33	CSN	**Canet-Saint-Nazaire**
4	ST	**Stagnone**	14	MAU	**Mauguio**	24	CO	**Comacchio**	34	BUT	**Butrinti**
5	MC	**Mar Chica**	15	FOG	**Fogliano**	25	BV	**Berre-Vaine**	35	KA	**Karavastas**
6	BAR	**Bardawil**	16	CA	**Caprolace**	26	OB	**Óbidos**	36	NA	**Narta-Valona**
7	TU	**Tunis**	17	BIG	**Biguglia**	27	SA	**St André**	37	BUD	**Buda**
8	MES	**Messolongui**	18	DI	**Diana**	28	EN	**Encañizada**	38	PIA	**Pialassa Baiona**
9	MAN	**Manzala**	19	UR	**Urbino**	29	TA	**Tancada**	39	MEL	**Mellah**
10	SC	**Scardovari**	20	FON	**Fondi**	30	AL	**Albufera**	40	BUR	**Burullus**

Fig. 1 The location of the lagoons considered in this study

Table 1 Bibliographical sources of information on lagoon features and fish assemblages and fishing yield, used in this study

Lagoons	Hydrographic parameters	Fish assemblage
Mar Menor	55, 56	5
Thau	6, 16, 26, 28, 60	46, 68
Venecia	10	10
Stagnone di Marsala	11, 12, 14	11
Mar Chica	8, 24	49
Bardawil	22, 58, 59	58
Tunis	60	7
Messolongui	45, 48, 66	44, 45
Manzala	1, 19, 20, 27	20, 65
Scardovari	36, 64	42, 43, 64
Burano	3	3
Prévost	4, 53	53
Mejean	53	53
Mauguio	23, 41	21, 38
Fogliano	40	50
Caprolace	40	50
Biguglia	17, 23	23
Diana	23, 61	23
Urbino	2, 23, 61	23
Fondi	3	3
Biban	47, 69	47, 69
Arnel	53	53
Orbetello	9	3, 9
Comacchio	37, 52	31

Table 1 continued

Lagoons	Hydrographic parameters	Fish assemblage
Berre-Vaine	62	
Óbidos	15, 25	15
Santo André	13	13
Encañizada	18	18
Tancada	18, 51	18
Albufera de Valencia	35, 39, 67, 70	71
Salses-Leucate	6, 29	29
Vaccarès	33, 63	57
Canet-Saint-Nazaire	30	30
Butrinti	54	54
Karavastas	54	54
Narta-Valona	54	54
Buda	18	18
Pialassa Baiona	34	34
Mellah	69	69
Burullus	69	69

(1) Abdel-Moati & Dowidar, 1990, (2) Agostini et al., 2003, (3) Ardizzone, 1984, (4) Bachelet et al., 2000, (5) Barcala, 1999, (6) Bec et al., 2003, (7) Ben Hassine, 1981, (8) Boussalwa et al., 2000, (9) Brando et al., 2004, (10) Busatto et al. 2002, (11) Calvo et al., 1986, (12) Campolmi et al., 1996, (13) Cancela da Fonseca et al., 1989, (14) Cavallaro et al., 1978, (15) Correia et al., 1997, (16) Crespi, 2002, (17) De Casabianca, 1975, (18) De Sostoa & De Sostoa, 1985, (19) Dowidar & Abdel-Moati, 1983, (20) Dowidar & Hamza, 1983, (21) Dumay et al., 2004, (22) Fouda et al., 1987, (23) Frisoni et al., 1983, (24) González & Conde, 1991, (25) Gordo & Cabral, 2001, (26) Gourret, 1897, (27) Halim & Guergues, 1981, (28) Hamon et al., 1979, (29) Hervé & Truslé, 1980, (30) Hervé & Truslé, 1981, (31–39) http://www.members.xoom.virgilio.it/passat/fauna.htm, http://www.sea.unep-wcmc.org/index.html, http://www.sea.unep-wcmc.org/sites/pa/0355v.htm~main, http://www.ambra.unibo.it/baiona/index.htm, http://www.clubdelamar.org/, http://www.deltadelpo.it/, http://www.dsa.unipr.it/lagunet/, http://www.educ-envir.org/~euziere/lagunes/lagune-2.html, http://www.sinix.net/paginas/albufera/index.htm, (40) Izzo et al., 2005, (41) Joyeux et al., 1992, (42) Kapetsky & Lasserre, 1984, (43) Kapetsky, 1984, (44) Katselis et al., 2003, (45) Kotsonias, 1984, (46) Lasserre, 1989, (47) Lemoalle & Vidy, 1984, (48) Leonardos & Sinis, 1997, (49) Lozano Cabo, 1953, (50) Mariani, 2001, (51) Menéndez et al., 2002, (52) Mistri et al., 2002, (53) Paris & Quignard, 1971, (54) Pejá et al., 1996, (55) Pérez Ruzafa et al., 2002, (56) Pérez Ruzafa et al., 2005, (57) Poizat et al., 2004, (58) Por & Ben-Tuvia, 1981, (59) Por, 1972, (60) Quignard, 1984, (61) Raimbault, 1972, (62) Riouall, 1977, (63) Rosecchi et al., 2001, (64) Rossi et al., 1984, (65) Rowntree, 1984, (66) Sioku-Frangou & Gotsis-Skretas, 1985, (67) Soria et al., 2002, (68) Tournier et al., 1983, (69) Vanden Bossche & Bernacsek, 1991, (70) Villena & Romo, 2003, (71) http://www.albufera.com

and 3), five of these (ACOM, LCOM, SCOM, ATRC, PMEDC) being measured for every existing inlet of each lagoon. Twenty-four of these variables are morphometric parameters or indexes related to the geomorphologic features of the lagoons (Chubarenko et al., 2005). Six variables deal with hydrographic characteristics and environmental variability and another six variables deal with water quality and productivity. Finally, the list of fish species for each lagoon, total species richness and fish catches were also recorded.

Unfortunately, the information available on lagoon characteristics is somewhat heterogeneous and incomplete and thus some lagoons were removed from the matrix according to the analyses performed. Variables such as lagoon-water renewal rates, which is related to species colonization (Pérez-Ruzafa & Marcos, 1992, 1993), are considered to be of great importance in explaining the biological structure and composition of lagoon biological assemblages. However, this parameter is unknown for most of the lagoons and cannot be used in this work. Nevertheless, it is related to other measured variables such as the openness or restriction ratio.

Analyses of data

We used Pearson's correlation coefficient to test the relationship between variables. Patterns in the geomorphologic and morphometrical characteristics of the lagoons and the composition of fish assemblages were explored using multivariate

Table 2 Morphometric, hydrographic and trophic variables considered in this study

Abbreviatons	Parameter description
Morphometric parameters	
SLAG	Lagoon surface (km^2)*
PERI	Lagoon perimeter (km)*
DMAX	Maximum diameter of the lagoon (m)*
DMIN	Minimum diameter of the lagoon (m)*
DPER	Perpendicular distance to the open sea coastline (m)*
DPAR	Parallel distance to the open sea coastline (m)*
PMAX	Maximum depth of the lagoon (m)
PMED	Mean depth of the lagoon (m)
VLAG	Lagoon volume (m^3)
CMAR	Number of inlets or channels
ACOM 2,3,4	$Inlet_{1,2,3,...}$width (m)*
LCOM 2,3,4	$Inlet_{1,2,3,...}$length (m)*
SCOM 2,3,4	Inlet surface (m)
ATRC (1,2,3,4)	Cross-sectional area of $channel_{1,2,3...}$ (m^2)
PMEDC(1,2,3,4)	Mean depth of $channel_{1,2,3...}$(m)
ACOMT	Total width of inlets (m)
LCOMT	Total length of inlets or channels (m)
SCOMT	Total surface of inlets/channels (m^2)
ATRCT	Total cross-sectional area of channels (m^2)
Hydrographic parameters	
TMXA	Maximum temperature of the water in the lagoon (°C)
TMNA	Minimum temperature of the water in the lagoon (°C)
SMAX	Maximum salinity of the water in the lagoon
SMIN	Minimum salinity of the water in the lagoon
DS	Salinity range or the difference between the maximum and minimum salinity within the lagoon.
DSM	Maximum absolute difference in salinity values between the lagoon and the open sea in the zone near the lagoon studied.
Trophic parameters	
NMAX	Maximum nitrate concentration in lagoon waters (µg-at/l).
NMIN	Minimum nitrate concentration in lagoon waters (µg-at/l).
PO4MAX	Maximum phosphate concentration in lagoon waters (µg-at/l).
PO4MIN	Minimum phosphate concentration in lagoon waters (µg-at/l).
CHLMAX	Maximum chlorophyll concentration in the lagoon (mg/m^3).
CHLMIN	Minimum chlorophyll concentration in the lagoon (mg/m^3).

analyses. A Principal Component Analyses (PCA) was performed for this purpose using square-root-transformed data (except in the case of the presence/absence of data on fish assemblages) and the CANOCO package (ter Braak, 1990). Rare species in the fish species matrix (present in one lagoon only) were downweighted following the CANOCO procedure in order to prevent their having an excessive influence on the ordination. In order to explore the relationship between the species matrix and the morphological and hydrological variables, a redundancy analysis (RDA) was performed. All calculations were performed on the covariance matrix, with square root transformed data. The relative contribution of each variable to the ordination established by the RDA was evaluated using a Monte Carlo permutation test after performing a forward selection of variables at a 0.1 level of significance. Furthermore, geomorphologic factors may act as synthetic variables for the underlying processes that determine the hydrological and trophic features, as is the case when studying the factors that might affect the fish assemblage composition in coastal lagoons. Therefore we separated both effects according to the methodology proposed by Borcard et al. (1992). Partitions of the variance were performed on an 18-cases matrix. By performing two canonical ordinations, each one constrained by a set of explanatory variables, one achieves the overall species importance value for the effects imposed by (1) the geomorphologic conditions and (2) the hydrographic characteristics (including trophic status). In the aforementioned analyses, the amount of variation in species data due to the influence of geomorphology on hydrographic conditions was extracted by both the hydrographic and geomorphologic sets of variables via partial canonical ordination (RDA) of the species and geomorphologic matrices while controlling for the effect of the environmental descriptors, and via the partial canonical ordination of the species and environmental data sets, controlling for geomorphology, using in both cases the other matrix as a covariable (Borcard et al., 1992).

We performed multiple regressions on all environmental variables (geomorphologic, hydro-

Table 3 Morphometric indexes of the lagoons considered in this study

Parameter	Formula	Description
Restriction ratio 1 (PRES)	$p_r = d/b$, $p_r = \sum d_i/b$	This restriction ratio is the ratio between the total width of lagoon entrances (d_i) and the parallel shore direction (b).
Restriction ratio 2 (PRES2)	$p_r = \sum d_i/P$	This restriction ratio is the ratio between the total width of lagoon entrances (d_i) and the lagoon perimeter (P).
Orientation or shape (POR)	$p_{or} = b^2/S_{lag} = S_{lag}/a^2$	The lagoon has orthogonal dimensions of the same order if $p_{or} \approx 1$. It is more elongated in a parallel (b) or perpendicular (a) sense to shore directions if $p_{or} \geq 1$ or $p_{or} \leq 1$, respectively. S_{lag} is the surface area of the lagoon.
Extreme depth (PDEP)	$P_{deep} = h_{max}/h_{av}$	The extreme depth parameter provides information on the deepest part of the lagoon and how it compares to mean depth.
Openness (PSEA)	$P_{sea} = \sum S_i^{in}/S_{lag}$	The openness parameter characterizes the potential influence of the sea on general lagoon hydrology, since entrance-way flow velocities are not included. Here, S_i^{in} is the cross-sectional area of i-th lagoon entrance for $i = 1$, n entrances, and S_{lag} is the lagoon surface area.
Shoreline development (PSH)	$P_{shore} = l\ (4\pi A)^{-0.5}$	The shoreline development parameter is the ratio of the length of shoreline (the lagoon perimeter (l)), to the circumference of the circle whose area (A) is equivalent to that of the lagoon.

logical and trophic), and their quadratic and cubic terms, for species richness and fishing catches in the framework of generalized linear models (GLM) (McCullagh & Nelder, 1989) with step-wise forward selection of variables (using $P < 0.05$ as the inclusion and/or rejection criterion).

Results

Atlanto-Mediterranean coastal lagoons show a high range of variability in size and hydrographic conditions, water quality, species richness and fishing yield (Table 4). The fish species richness in the lagoons studied ranged between 6 and 48, with a mean of 23.4 species. Although a total of 179 fish species were inventoried in the lagoons we studied, only 98 of them are common to at least two lagoons (Table 5). Six species (*Anguilla anguilla*, *Atherina boyeri*, *Dicentrarchus labrax*, *Liza ramada*, *Mugil cephalus* and *Sparus auratus*) were very frequent, being present in more than 70% of the lagoons studied. Those families represented by a greater number of very frequent or frequent species were: Mugilidae (5 species); and Sparidae (3 species). On the other hand, 121 species could be considered occasional or specific to a few lagoons, since they were present in less than 15% of the cases studied.

From a geomorphologic point of view, most of the variability in lagoon morphology relates to size, lagoon volume (VLAG), surface (SLAG) and diameter (DMAX). The first axis of the PCA for the geomorphologic variables of the lagoons accounts for 99.8% of the total variance in data and is mainly related to these three variables (Fig. 2). The second component only accounts for an additional 0.2% and is related to the total surface (SCOMT) and length (LCOMT) of the inlets, and to a lesser extent to the shoreline development parameter (PSH: the ratio of the length of shoreline, the lagoon perimeter, and the circumference of the circle whose area is equivalent to that of the lagoon).

When performing the RDA on the lagoon species matrix using geomorphologic parameters as explanatory variables, the sum of all canonical eigenvalues reached 60.7% of the variation and the first two axis explained 33.9% of the variance in the composition of lagoon fish assemblages. The more important factors were the total transversal surface of inlets (ATRCT), the lagoon perimeter (PERI), the restriction parameter (PRES2) and several variables related with lagoon size, such as maximum depth (PMAX) and lagoon diameter (DMAX) and volume (VLAG) (Fig. 3a).

Table 4 Maximum, minimum and mean values and mean standard error for all variables considered in this study, indicating the number (*n*) of lagoons for which data were reported (see Tables 2 and 3 for the abbreviations of variables)

Parameters	Units	*n*	Max	Min	Mean	SE
SLAG	km^2	39	1168.36	1.25	106.08	2.441
PERI	km	38	1236.42	7.28	92.80	2.347
DMAX	m	39	57000.00	2378.78	12484.57	19.180
DMIN	m	39	14000.00	172.41	2381.24	8.701
DPER	m	39	30267.33	532.73	5587.18	12.267
DPAR	m	39	72565.11	1368.29	13083.47	20.765
PMAX	m	38	35.00	0.60	5.35	0.448
PMED	m	39	20.00	0.35	2.30	0.302
VLAG	m^3	39	1.46E + 09	7.55E + 05	1.92E + 08	3103.538
CMAR		38	4	0	1.47	0.158
ACOM	m	38	2661.71	0.00	338.96	4.253
LCOM	m	38	8778.93	0.00	1403.03	7.184
SCOM	m^2	38	3268659.01	0.00	197382.67	124.059
ATRC1	m^2	33	7371.91	0.00	762.40	7.053
PMEDC1	m	33	9.00	0.00	1.83	0.243
ACOMT	m	38	7015.41	0.00	550.40	5.854
LCOMT	m	38	17787.53	0.00	2245.48	9.584
SCOMT	m^2	38	5244335.07	0.00	314396.69	159.590
ATRCT	m^2	38	16244.42	0.00	1023.24	8.654
PRES		38	0.56	0.00	0.05	0.055
PRES2		38	0.08	0.00	0.01	0.023
POR		39	21.23	0.30	3.55	0.298
PDEP		38	14.00	1.12	2.53	0.245
PSEA		33	0.00	0.00	0.00	0.001
PSH		38	15.88	1.32	3.00	0.284
TMXA	°C	35	32.00	20.00	27.81	0.264
TMNA	°C	34	16.00	3.00	8.76	0.333
SMAX		36	78.90	3.40	38.19	0.617
SMIN		35	43.70	1.90	19.25	0.603
DS		35	41.00	0.95	18.12	0.555
DSM		34	36.61	6.10	20.96	0.555
NMAX	μM/l	25	400.00	0.23	44.81	1.872
NMIN	μM/l	24	390.00	0.00	17.51	1.843
PO4MAX	μM/l	28	29.68	0.03	3.97	0.464
PO4MIN	μM/l	25	9.00	0.00	0.56	0.266
CHLMAX	μg/l	29	593.00	0.21	70.30	2.187
CHLMIN	μg/l	26	154.30	0.00	8.40	1.074
Fishing Yield	kg/ha year	27	260.00	6.00	86.08	1.663
Species Richness	No. of spp	39	48	6	23.41	0.563

In the case of the RDA performed on the species matrix using trophic parameters as explanatory variables, these (mainly the chlorophyll *a* concentration (CHLMAX) and the maximum phosphate concentration (PO4MAX)) explained 34.1% of the variability in the composition of fish assemblages, determining a trophic gradient along the first axis, with eutrophic conditions in the negative part (Fig. 3b). The second axis explains an additional 24.6% and is determined by the maximum nitrate concentration in the positive part, and the highest values for the minimum phosphate concentration in the negative part (Fig. 3b). Therefore, the first quadrant would include mesotrophic and oligotrophic situations wherein production is limited by the phosphate concentration, and the negative part of the second axis involves situations in which the nitrate concentration is the limiting factor.

Finally, when using hydrographic variables as constraints for the RDA, the first two axes explained 59.7% of the variance in fish assemblages (Fig. 3c). The first component (42.7% of the variance) determined a gradient, from hy-

Table 5 Fish species present in the lagoons studied and their frequency

Family	Species	f
Anguillidae	*Anguilla anguilla*	89.2
Congridae	*Conger conger*	5.4
Belonidae	*Belone belone*	32.4
Hemiramphidae	*Hemiramphus* sp*	2.7
Syngnathidae	*Hippocampus hippocampus*	8.1
	Hippocampus ramulosus	18.9
	Nerophis ophidion	5.4
	Syngnathus abaster	51.4
	Syngnathus acus	13.5
	Syngnathus typhle	13.5
Cyprinodontidae	*Aphanius fasciatus*	35.1
	*Aphanius dispar**	2.7
	Lebias ibera	10.8
Poeciliidae	*Gambusia affinis*	21.6
Caproidae	*Capros aper**	2.7
Atherinidae	*Atherina boyeri*	73.0
	Atherina hepsetus	8.1
	*Atherina presbyter**	2.7
Ammodytidae	*Ammodytes tobianus**	2.7
Blenniidae	*Aidablennius sphynx**	2.7
	*Coryphoblennius galerita**	2.7
	*Lipophrys canevae**	2.7
	*Lipophrys dalmatinus**	2.7
	*Parablennius gattorugine**	2.7
	*Parablennius sanguinolentus**	2.7
	*Parablennius incognitos**	2.7
	Parablennius tentacularis	5.4
	*Paralipophrys trigloides**	2.7
	Salaria pavo	29.7
	Salaria fluviatilis	8.1
Bramidae	*Brama brama*	8.1
Callanthidae	*Callanthias rubber**	2.7
Callionymidae	*Callionymus risso**	2.7
	*Callionymus lyra**	2.7
	*Callionymus pusillus**	2.7
	*Callionymus reticulates**	2.7
	*Callionymus maculates**	2.7
Carangidae	*Lichia amia*	16.2
	Seriola dumerili	5.4
	*Trachurus trachurus**	2.7
	Trachurus mediterraneus	5.4
Centrarchidae	*Lepomis gibbosus*	8.1
	Micropterus salmoides	13.5
Cichlidae	*Haplochromis* sp*	2.7
	Hemichromis sp*	2.7
	*Oreochromis upembae**	2.7
	*Oreochromis aureus**	2.7
	*Sarotherodon galilaeus**	2.7
	*Tilapia zillii**	2.7
Clinidae	*Clinitrachus argentatus**	2.7
Gobiidea	*Aphia minuta**	2.7
	*Deltentosteus quadrimaculatus**	2.7
	Gobius cobitis	8.1
	Gobius cruentatus	5.4
	*Gobius geniporus**	2.7
	Gobius niger	37.8
	Gobius paganellus	10.8
	Gobius bucchichi	8.1
	*Gobius ater**	2.7
	Knipowitschia panizzai	5.4
	*Millerigobius macrocephalus**	2.7
	*Pomatoschistus tortonesei**	2.7
	Pomatoschistus minutus	29.7
	Pomatoschistus microps	29.7
	Pomatoschistus canestrini	5.4
	Pomatoschistus marmoratus	16.2
	*Pomatoschistus knerii**	2.7
	*Pomatoschistus pictus**	2.7
	*Silhouettea aegyptia**	2.7
	*Zebrus zebrus**	2.7
	Zoosterisessor ophiocephlus	10.8
Labridae	*Coris julis**	2.7
	*Ctenolabrus rupestris**	2.7
	Labrus viridis	5.4
	*Labrus merula**	2.7
	Labrus bergylta	5.4
	Symphodus cinereus	13.5
	Symphodus ocellatus	5.4
	Symphodus roissali	5.4
	Symphodus tinca	5.4
	Symphodus melops	5.4
	*Symphodus mediterraneus**	2.7
	Symphodus bailloni	5.4
	Thalassoma pavo	5.4
Moronidae	*Dicentrarchus labrax*	83.8
	*Dicentrarchus punctatus**	2.7
Mugilidae	*Chelon labrosus*	67.6
	Liza aurata	67.6
	Liza ramada	73.0
	Liza saliens	59.5
	*Liza carinata**	2.7
	Mugil cephalus	78.4
	Oedalechilus labeo	21.6
Mullidae	*Mullus barbatus*	27.0
	Mullus surmuletus	37.8
Percidae	*Sander lucioperca**	2.7
Pomacentridae	*Chromis chromis**	2.7
Pomatomidae	*Pomatomus saltatrix*	13.5
Sciaenidae	*Argyrosomus regius*	13.5
	*Sciaena umbra**	2.7
	Umbrina cirrosa	16.2
Scombridae	*Scomber scombrus**	2.7
Serranidae	*Epinephelus marginatus**	2.7
	Epinephelus aeneus	10.8
	*Serranus scriba**	2.7

Table 5 continued

Family	Species	f
Sparidae	*Boops boops*	16.2
	Crenidens crenidens	5.4
	Dentex dentex	8.1
	Diplodus annularis	54.1
	Diplodus sargus	37.8
	Diplodus vulgaris	35.1
	Diplodus puntazzo	27.0
	Diplodus bellottii	8.1
	Diplodus cervinus	10.8
	Lithognathus mormyrus	43.2
	Oblada melanura	8.1
	*Pagellus acarne**	2.7
	*Pagellus bogaraveo**	2.7
	Pagrus pagrus	5.4
	Pagrus coeruleostictus	8.1
	Sarpa salpa	27.0
	Sparus aurata	75.7
	Spondyliosoma cantharus	8.1
Sphyraenidae	*Sphyraena sphyraena*	0.0
Trachinidae	*Echiichthys vipera**	2.7
	Trachinus sp*	2.7
Tripterygiidae	*Tripterygion tripteronotus**	2.7
	Tripterygion spp*	2.7
Bothidae	*Arnoglossus kessleri**	2.7
	Arnoglossus laterna	13.5
	*Arnoglossus imperialis**	2.7
	*Bothus podas**	2.7
Pleuronectidae	*Platichthys flesus*	32.4
	*Pleuronectes platessa**	2.7
Scophthalmidae	*Scophthalmus maximus*	10.8
	Scophthalmus rhombus	10.8
Soleidae	*Buglossidium luteum**	2.7
	*Dicologlossa cuneata**	2.7
	*Microchirus ocellatus**	2.7
Soleidae	*Microchirus theophila*	5.4
	*Solea impar**	2.7
	Solea solea	64.9
	Solea senegalensis	8.1
	Solea lascaris	16.2
Clupeidae	*Alosa pontica**	2.7
	Alosa alosa	10.8
	Alosa fallax	16.2
	Sardina pilchardus	32.4
	*Sprattus sprattus**	2.7
Engraulidae	*Engraulis encrasicolus*	35.1
Gasterosteidae	*Gasterosteus aculeatus*	13.5
Salmonidae	*Salmo trutta*	5.4
	*Scyliorhinus canicula**	2.7
	*Scyliorhinus stellaris**	2.7
Triakidae	*Mustelus mustelus**	2.7
Scorpaenidae	*Scorpaena porcus*	16.2
Triglidae	*Chelidonichthys lucerna*	10.8
	*Trigla lyra**	2.7
Balistidae	*Balistes carolinensis**	2.7
Cobitidae	*Cobitis taenia**	2.7

Table 5 continued

Family	Species	f
Cyprinidae	*Abramis brama**	2.7
	*Alburnus alburnus**	2.7
	*Barbus barbus**	2.7
	Barbus graellsi	8.1
	*Blicca bjoerkna**	2.7
	Carassius carassius	5.4
	Carassius auratus	13.5
	Chondrostoma toxostoma	10.8
	Cyprinus carpio	21.6
	Gobio gobio	8.1
	*Labeo niloticus**	2.7
	*Pseudorasbora parva**	2.7
	Vimba melanops	5.4
Rajidae	*Raja asterias*	13.5
	Raja spp*	2.7
Bagridae	*Bagrus bajad**	2.7
Ictaluridae	*Ameiurus nebulosus*	5.4
	*Ameiurus melas**	2.7
Siluridae	*Silurus glanis**	2.7
Petromyzontidae	*Petromyzon marinus*	10.8

Species reported in only one lagoon are indicated by *

persaline conditions in the negative part, to situations with a fresh water influence and greater differences from the open sea (DSM) and between different areas or seasons in the same lagoon (DS) in the positive part. The second axis was determined by a temperature gradient with the highest temperatures in the positive part.

Figure 4a shows the ordination of the lagoons according to their fish assemblages and the constraining variables when all variables, geomorphologic, hydrographic and trophic, are considered simultaneously via RDA. These variables are plotted in Fig. 4b. The first quadrant is determined by the total transversal surface of inlets (ATRCT), the lagoon perimeter (PERI), the restriction parameter (PRES2) and maximum depth (PMAX). Other parameters related to lagoon size, such as the lagoon volume or maximum diameter (VLAG, DMAX), became less relevant. Appearing in this quadrant, related to the positive part of the second axis, are shoreline development (PSH) and the highest minimum concentrations for phosphate (PO4MIN) and nitrate (NMIN). In the opposite quadrant, weighting on the negative part of axis

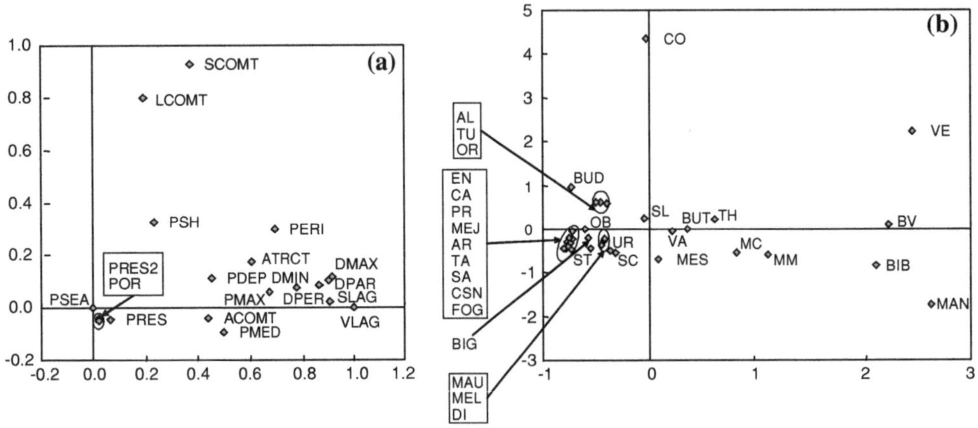

Fig. 2 Ordination diagrams for the Principal Component Analysis performed on geomorphologic lagoon characteristics showing (**a**) the variables and (**b**) the lagoons (see Tables 1 and 2 for the abbreviations of variables and Fig. 1 for lagoons)

one and two, the highest minimum and maximum chlorophyll *a* concentrations (CHLMIN, CHLMAX) and maximum nitrate (NMAX) and phosphate (PO4MAX) concentrations appear, thus determining eutrophic conditions in lagoons.

Partitioning the variation

However, as expected, hydrographic and trophic factors showed a strong dependence on geomorphologic features. Maximum temperature, the range in salinity inside the lagoon and the difference in salinity between the lagoon and the open sea are positively correlated with restriction parameters (PRES, PRES2) and, in general, nutrient concentration and phytoplanktonic productivity were negatively correlated with maximum salinity, indicating their dependence on fresh water inputs and/or isolation from the sea. In the same context, the minimum concentration of nitrates was positively correlated with the lagoon perimeter ($r = 0.88$, $P < 0.0001$) and shoreline development (Table 6). Therefore, a partition of the variance was performed in order to separate hydrographic and geomorphologic features.

The four steps according to the Borcard et al. (1992) analytical procedure, using RDA, gave the following results:

(1) The RDA on the species matrix, constrained by the environmental (hydrographic and trophic variables) matrix gave 77.8% as the sum of all canonical eigenvalues.
(2) The sum of all canonical eigenvalues for the RDA on the species matrix, when constrained by geomorphologic features was 100.2%
(3) The RDA on the species matrix, constrained by the environmental (hydrographic and trophic variables) matrix, after removing the effect of geomorphologic variables gave 3.2% as the sum of all canonical eigenvalues.
(4) The sum of all canonical eigenvalues for the RDA on the species matrix, when constrained by the extended matrix of geomorphologic features, after removing the effect of hydrographic variables was 22.2%

In the CANOCO version of the RDA, the sum of eigenvalues can be read directly as fractions of explained variation. Thus, the total explained variation came to:

$$77.8 + 22.2 = 100\%$$

The covariation between hydrographic variables and geomorphologic features was 77.8–3.2 = 74.6% of the total variation.

The variation due exclusively to the hydrographic conditions of the lagoons was only 3.2%,

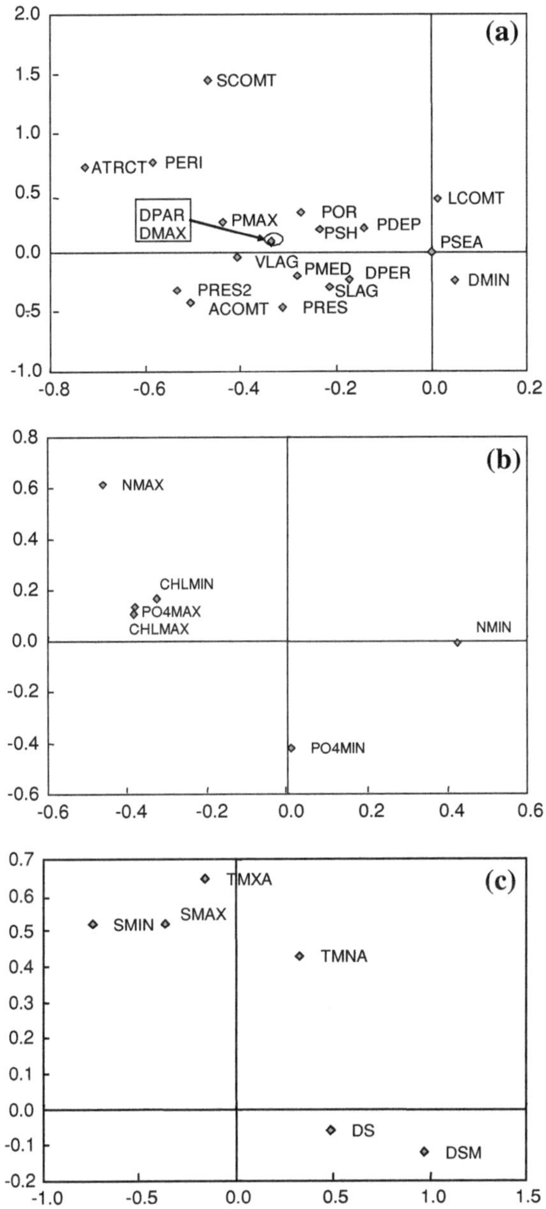

Fig. 3 Ordination diagrams of the explanatory variables for the fish assemblage composition of Atlanto-Mediterranean coastal lagoons in the RDA performed using: (**a**) geomorphologic lagoon characteristics; (**b**) variables indicating the trophic status of the water column; and (**c**) hydrographic variables as constraining variables (see Tables 1 and 2 for abbreviations)

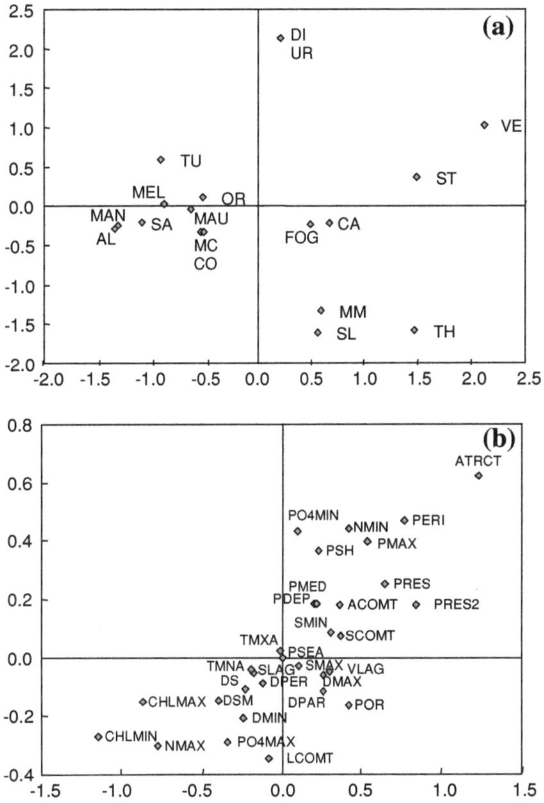

Fig. 4 Ordination diagrams (**a**) for lagoons and (**b**) for the explanatory variables of the fish assemblage composition of Atlanto-Mediterranean coastal lagoons in the RDA performed using all the variables together as constraining variables (see Tables 1 and 2 for abbreviations)

while the variation explained by geomorphologic features that do not coincide with hydrographic variables, reached 22.2%.

Multiple regression analyses

Factors related to species richness in coastal lagoons

Aside from the faunistic composition of the assemblages, the number of fish species present in a lagoon is partially explained by a few variables (Fig. 5). The number of species in a lagoon is negatively related to the maximum difference in salinity between lagoon waters and the sea (adj. $R^2 = 0.24$, $P < 0.005$) (Fig. 5a) and to the maximum chlorophyll *a* concentration (adj. $R^2 = 0.28$, $P < 0.005$) (Fig. 5b). In the context of generalized linear models (GLM) the best subset of two variables explaining species richness includes the total transversal area of the inlets,

Table 6 Pearson's correlation (Pearson's $|r| > 0.5$) and significance level ($P < 0.05$) between the hydrographic, trophic and geomorphologic characteristics of the lagoons

	SLAG	PERI	DMAX	DMIN	PMAX	PMED	VLAG	ACOM	LCOM	SCOM	ATR1	PMD1	ACOT
TMXA													
P-value													
TMNA										–0.52			
P										0.04			
PSEA												0.52	
P												0.04	
PSH		0.80			0.52						0.69	0.88	
P		0.00			0.04						0.00	0.00	
SMIN													
P													
DS	0.63			0.64				0.67					0.64
P	0.01			0.01				0.00					0.01
DSM				0.53									
P				0.03									
PROD													
P													
NMAX									0.56				
P									0.02				
NMIN		0.88	0.66		0.96		0.59				0.90	0.72	
P		0.00	0.01		0.00		0.02				0.00	0.00	
PO4MAX													
P													
PO4MIN						0.60							
P						0.01							
CHLMAX									0.54				
P									0.03				
CHLMIN									0.55				
P									0.03				

which communicate with the sea (ATRCT) and DSM parameters (adj. $R^2 = 0.34$, $P < 0.05$) (Table 7, Fig. 5c).

The general model for explaining lagoon fish species richness, when all factors are considered in the regression analyses using stepwise selection of variables, includes: a positive linear relationship with the PSEA index or openness parameter, which characterizes the potential influence of the sea on general lagoon hydrology; a negative relationship with the cubic term of the minimum phosphate concentration in lagoon waters (PO4MIN); and the linear and cubic term of the lagoon volume (VLAG) (adjusted $R^2 = 0.75$, $P < 0.05$) (Table 8). Therefore, the number of fish species in a lagoon increases with the influence of the sea on lagoon hydrology and with the size of the lagoon (as the negative effect of the cubic term of the lagoon volume is very low due to its coefficient). Otherwise, the species richness in a lagoon is strongly limited by the phosphate concentration in the water column.

Factors related to fishing yield

On the other hand, fishing yield, expressed as the annual fishing catch per lagoon surface unit (kg/Ha*year), is related to several hydrographic and geomorphologic parameters. Catches rise with increasing total length of inlets (adjusted $R^2 = 0.19$, $P = 0.01$) (Fig. 6a) and decrease with increasing mean depth of the lagoon (adjusted $R^2 = 0.18$, $P < 0.05$) (Fig. 6b). Fishing yield, is also positively related to the chlorophyll *a* concentration (Fig. 7a) (adjusted $R^2 = 0.43$, $P = 0.000$), and tends to diminish with increasing

SCOT	ATRCT	PRES	PRES2	PDEP	PSEA	PSH	SMAX	SMIN	DSM	NMAX	PO4MAX	CHLMAX
					0.52							
					0.04							
–0.53												
0.03												
	0.67			0.52	0.52							
	0.00			0.04	0.04							
							0.55					
							0.03					
		0.59	0.53					–0.55	0.56			
		0.02	0.04					0.03	0.02			
		0.52						–0.90				
		0.04						0.00				
							–0.53	–0.52				0.52
							0.04	0.04				0.04
							–0.79		0.59			
							0.00		0.02			
	0.91			0.80		0.61						
	0.00			0.00		0.01						
							–0.68			0.82		
							0.00			0.00		
							–0.76	–0.53	0.59	0.93	0.71	
							0.00	0.03	0.02	0.00	0.00	
							–0.78			0.92	0.73	0.98
							0.00			0.00	0.00	0.00

maximum salinity (adjusted $R^2 = 0.30$; $P < 0.05$) and, to a lesser extent, with the minimum phosphate concentration in lagoon waters (adjusted $R^2 = 0.18$, $P < 0.05$) (Fig. 7b, c, respectively). Despite the large leverage in the case of Albufera (Fig. 7a), the regression equation remains nearly identical when this case is removed from the analyses.

On including all parameters (morphological, hydrographic and trophic) in the GLM regression analyses with stepwise selection of variables, the variables selected by the model are the cubic term of coastal development (PSH) and the linear term of the maximum chlorophyll concentration in the water column (adjusted $R^2 = 0.44$, $P < 0.05$) (Table 9). Therefore, the fishing yield increases linearly with the maximum chlorophyll concentration in the lagoon and exponentially with coastal development (Fig. 8).

Discussion

Most of the physical and environmental variability in Atlanto-Mediterranean coastal lagoons is related to lagoon size, differences in salinity with respect to the open sea and the trophic status of the water column. The fish assemblages, in terms of species richness and composition, are best explained by the degree of communication with the open sea, expressed as the total area of the transversal section of inlets (ATRCT) or as the openness parameter (PSEA), the size of the lagoon (volume and perimeter) and the trophic status of the water column (chlorophyll *a* concentration and minimum nitrate and phosphate concentrations), as well as other hydrographic factors such as minimum salinity, freshwater influence and minimum temperature. However, as stated before, hydrographic and trophic factors showed a strong dependence on geomorphologic features and a

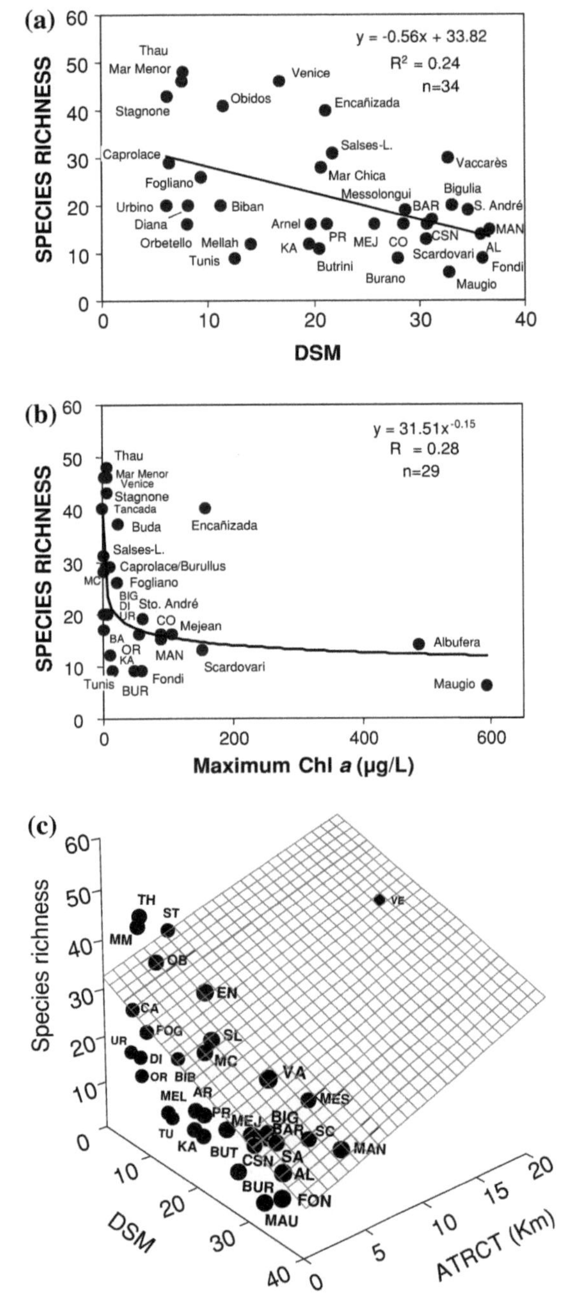

Fig. 5 The relationship between the number of fish species in a coastal lagoon and (**a**) the difference in salinity between lagoon waters and the sea ($P < 0.005$), (**b**) the maximum chlorophyll *a* concentration the lagoon water column ($P < 0.005$) and (**c**) the two parameters (DSM or difference in salinity between lagoon waters and the sea and ATRCT or total transversal area of the inlets connecting the lagoon and the open sea) selected in the multiple regression analyses with forward selection as the best subset for two variables ($P < 0.05$)

Table 7 Best subset of two variables in a GLM context explaining lagoon fish species richness, while considering the morphometric, hydrological and trophic descriptors of the lagoons (Cases included = 19; adj. $R^2 = 0.34$)

Resulting variable	Stepwise model coefficient	Std error	*P*
Constant	32.6750	4.75127	0.0000
DSM	–0.58680	0.22372	0.0185
ATRCT	0.00146	6.715E-04	0.0454

(See Table 1 for the abbreviations of variables)

Table 8 General model explaining lagoon fish species richness, considering the morphometric, hydrological and trophic descriptors of the lagoons in the regression analyses with stepwise selection of variables (Cases included = 19; adj. $R^2 = 0.75$)

Resulting variable	Stepwise model coefficient	Std error	*P*
Constant	14.8697	2.10334	0.0000
$PO4MIN^3$	–71.1669	27.4620	0.0213
PSEA	177983	36065.9	0.0002
VLAG	8.989E-08	1.323E-08	0.0000
$VLAG^3$	–4.158E-26	6.672E-27	0.0000

(See Tables 1 and 2 for the abbreviations of variables, $PO4MIN^3$ and $VLAG^3$ are the cubical values of PO4MIN and VLAG)

large lagoon perimeter with shoreline development favours nutrient input. According to these findings, fishing yield increases with the Pshore (PSH) parameter (Fig. 8).

Otherwise, fishing catches decrease with increasing mean depth of the lagoon (Fig. 6b). In fact, in shallow lagoons, bottoms are usually well irradiated, currents and hydrodynamics are closely conditioned by bottom topography and wind affects the entire water column promoting resuspension of materials, nutrients and small organisms from the sediment surface layer, thus increasing productivity (Conde et al., 1999; Gamito et al., 2005; Miller et al., 1990; Nixon, 1982; Suzuki et al., 2002).

On the other hand, species richness, which did not show a significant Pearson's correlation with any hydrographic or geomorphologic parameter at $P < 0.05$, were positively related to the total transversal area of inlets (Fig. 5c) and the openness parameter (Table 8) and negatively related to the absolute difference in salinity between lagoon

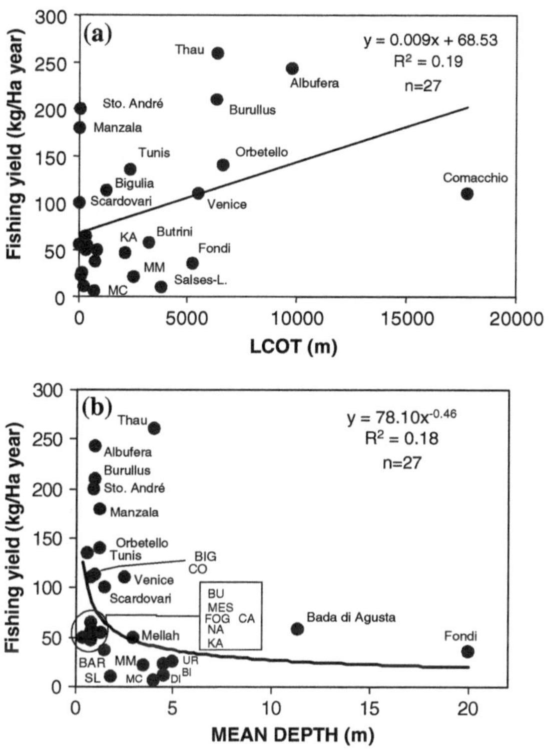

Fig. 6 The relationship between the fishing yield in a coastal lagoon and (**a**) the total length of the inlets connecting the lagoon to the open sea ($P = 0.01$); and (**b**) the mean depth of the lagoon ($P < 0.05$)

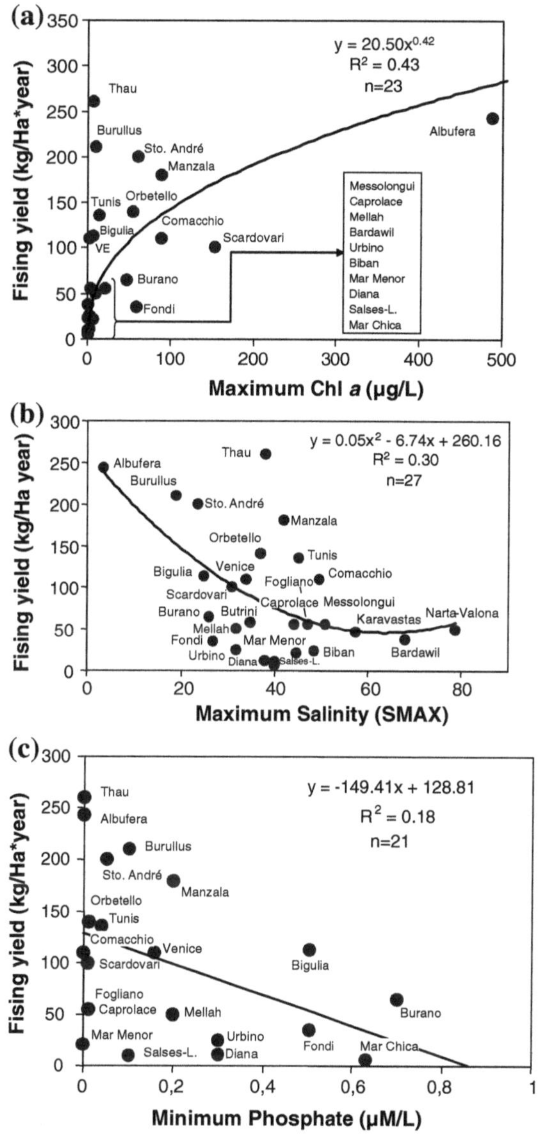

Fig. 7 The relationship between the fishing yield in a coastal lagoon and (**a**) the maximum chlorophyll *a* concentration in the lagoon water column ($P = 0.000$), (**b**) the maximum salinity of lagoon waters ($P < 0.05$) and (**c**) the minimum phosphate concentration in the lagoon water ($P < 0.05$)

waters and the sea (Fig. 5a, c; Table 7). This agrees with the confinement theory by Guelorget & Perthuisot (1983) reformulated by Pérez-Ruzafa & Marcos (1992, 1993) and reinforces the idea that fish species richness in coastal lagoons is determined by colonization rates from the open coastal sea (Mariani, 2001; Pérez-Ruzafa et al., 2004, 2006) and, depending on the estuarine influence, by colonization rates from rivers. Therefore, from restricted to leaky lagoon continuums, sensu Kjerfve (1994), under increasing renewal rates and decreasing water residence times, the penetration of species by means of trophic or reproductive migratory activities in swimming stages or randomless ones in pelagic and planktonic stages such as eggs, larvae or juveniles, is facilitated.

The positive relationship between species richness and lagoon volume, a sinthetic expresion of surface and depth, is coherent with the expectation that larger lagoons could provide a greater diversity of environments and types of bottoms with specific assemblages (Pérez-Ruzafa, 1989; Pérez-Ruzafa et al., 2006). However, it is currently difficult to obtain detailed information on the percentage of recovery of different kinds of substrata and benthic assemblages in a set of coastal lagoons in order to test and quantify the real influence of habitat heterogeneity.

Table 9 General model explaining lagoon fishing yield, while considering the morphometric, hydrological and trophic descriptors of the lagoons in the regression analyses with stepwise selection of variables (Cases included = 17; adj. R^2 = 0.44)

Effect	Coefficient	Std error	*P*
Constant	35.538	10.334	0.004
CHLMAX	0.538	0.217	0.027
PSH^3	0.022	0.007	0.006

(See Tables 1 and 2 for the abbreviations of variables. PSH3 is the cubic value of PSH)

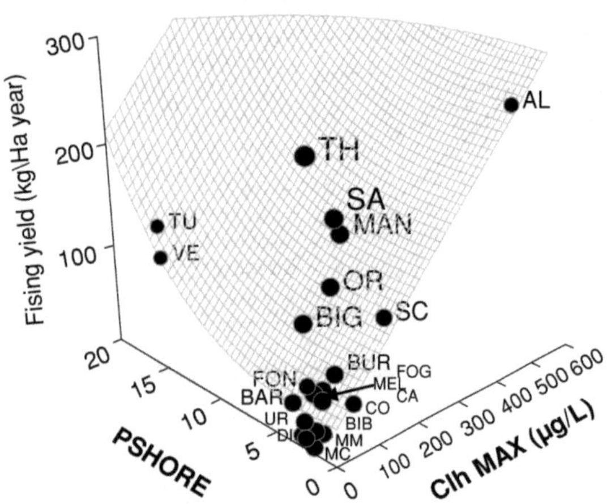

Fig. 8 Best regression model (GLM) after the forward selection of variables, explaining the relationship between the fishing yield in a coastal lagoon and their geomorphologic and hydrographic characteristics. The variables selected for the model were PSH or shoreline development ($p < 0.01$) and the maximum chlorophyll concentration in the lagoon water column ($p < 0.05$)

Finally, the positive influence of the maximum chlorophyll *a* concentration on fishing yield and the negative influence on species richness suggest that the increase in primary productivity favours the abundance of some species at the expense of species richness. This could agree with ecological theory and the expected response of ecosystems and community structures, in terms of disminishing ecological diversity, when face with high external nutrients and energy inputs (Tilman, 1982, 1999).

In summary, geomorphologic features alone explained 22% of the variance in the canonical analyses and an additional 75% in conjunction with the hydrographic and trophic characteristics of the lagoon, the latter two explaining only 3% of the fish assemblage composition. Geomorphologic parameters (mainly volume, sea influence, and shoreline development) are easy to measure and can be used as the basis for typification as required in the context of the European *Water Framework Directive-WFD*. Furthermore, all of these parameters are strongly affected by changes in sea level, coastal engineering and other human activities. A rise in sea level involves an increase in lagoon size and depth and in some areas can affect the isolation status with respect to the open sea. Human engineering modifies shoreline development, sedimentation rates and depth via land reclamation or building dykes or marinas, as well as affecting the influence of the open sea, modifying the structure of natural inlets or building new ones. Some of these activities could be intentionally directed at improving biological features. However, the effects on species richness can be the direct opposite of those on fishing yield or species composition and therefore, when designing management strategies and evaluating the impact of human activities, it would be necessary to consider the importance of maintaining the naturalness of these exclusive ecosystems and not only the improvement of one particular characteristic.

Acknowledgements This study was financed by the project CGL2004-06891 (National R&D Program, Ministry for Education and Science, Spain).

References

Abdel-Moati, A. R. & N. M. Dowidar, 1990. Salinity and major ions in Lake Manzala, Egypt. Rapp Comm int Mer Medit 32 fasc 1: 68.

Agostini, S., B. Marchand & G. Pergent, 2003. Temporal and spatial changes of seagrass meadows in a Mediterranean coastal lagoon. Oceanologia Acta 25: 297–302.

Allongi, D. M, 1998. Coastal Ecosystem Processes. CRC Press, Boca Raton.

Anonymous, 1959. Final resolution. The Venice System for the Classification of Marine Waters according to salinity. Archivio de Oceanografia e Limnologia 11(suppl): 243–245.

Ardizzone, G. D., 1984. Biological characteristics, exploitation and management of fishery resources in coastal lakes of central Italy. FAO Studies and Reviews 61(2): 461–476.

Bachelet, G., X. Montaudouin, I. Auby & P. J. Labourg, 2000. Seasonal changes in macrophyte and macrozoobenthos assemblages in three coastal lagoons under varying degrees of eutrophication. ICES Journal of Marine Science 57: 1495–1506.

Barcala, E., 1999. Estudio ecológico de la ictiofauna bentónica del Mar Menor. Unpublished PhD. Universidad de Murcia.

Barnes, R. S. K., 1980. Coastal Lagoons. Cambridge University Press, Cambridge.

Bec, B., Y. Collos, B. Bibent, B. Mostajir, V. Pons, P. Souchu & A. Vaquer, 2003. Limiting nutrients of pico and nanophytoplankton in two mediterranean coastal lagoons with different trophic status. In Murray, C. N., J. M. Zaldívar, G. Giordani & P. Viaroli (eds), Proceedings of the International Conference on Southern European Coastal Lagoons: The Influence of River Basin-Coastal Zone interactions Ferrara. Italy: 32.

Ben Hassine, O. K., 1981. Etude de l'infestation des poissons (autres que les muges) par les copepodes parasites dans les lacs de Tunis et de l'Ischkeul. Rapp Comm int Mer Medit 27: 125–129.

Bianchi, C. N., 1988 Caratterizzazione bionomica delle lagune costiere italiane. Acqua and Aria 4: 15–20.

Borcard, D., P. Legendre & P. Drapeau, 1992. Partialling out the spatial component of ecological variation. Ecology 73: 1045–1055.

Boussalwa, E. H., A. Douira & N. Mokhtar, 2000 Contribution à l'étude typologique d'une lagune exposée à la pollution, la lagune de Nador (Maroc): distribution des crustacés. Cahiers de Biologie Marine 41: 255–263.

Brando, V. E., R. Ceccarelli, S. Libralato & G. Ravagnan, 2004. Assessment of environmental management effects in a shallow water basin using mass-balance models. Ecological Modelling 172: 213–232.

Busatto, T., P. Franzoi, P. F. Ghetti, G. Maio, E. Marconato, L. Mizzan, M. Pellizzato, P. Penzo & S. Salviati, 2002. Pesci molluschi e crostacei della laguna di Venecia. Risorse ittiche e ambiente lagunare tra storia e innovazione. Ed. Cicero, Venecia.

Calvo, S., G. Genchi, A. Lugaro, C. Frada'Orestano, R. Barone & F. Di Bernardo, 1986. Osservazioni ecologique su una laguna siciliana (Lo Stagnone, Trapani):Nutrienti, clorofilla e parametri batteriologici. Atti del VII Congreso AIOL 11: 185–193.

Campolmi, M., P. Franzoi & A. Mazzola, 1996. Observations on pipefish (Syngnathidae) biology in the Stagnone lagoon (west Sicily). Publicaciones Especiales del Instituto Español de Oceanografía 21: 205–209.

Cancela da Fonseca, L., A. M., Costa & J. M Bernardo, 1989. Seasonal variation of benthic and fish communities in a shallow land-locked coastal lagoon (St.André, SW Portugal). Topics in Marine Biology (Scientia Marina) 53: 663–669.

Cavallaro, G., P. Arena, A. Cefali, A. Potoschi, P. Sorbilli & A. Di Natale, 1978. Studio di un ambiente lagunare:lo Stagnone di Marsala. Ed. Delphinus, Augusta.

Chubarenko, B., V. G. Koutitonsky, R. Neves & G. Umgiesser, 2005. Modeling concepts. In Gönenç, I. E. & J. P. Wolflin (eds), Coastal Lagoons: Ecosystem Processes and Modeling for Sustainable Use and Development. CRC Press, Boca Ratón: 231–306.

Clark, J. R., 1998. Coastal Seas, The Conservation Challenge. Blackwell Science, Oxford.

Conde, D., S. Bonilla, L. Aubriot, R. de León & W. Pintos, 1999. Comparison of the areal amount of chlorophyll *a* of planktonic and attached microalgae in a shallow coastal lagoon. Hydrobiologia 408/409: 285–291.

Correia, M. J., M. J. Costa & L. S. Gordo, 1997. Trophic groups of fish in the Óbidos lagoon (Portugal). Publicaciones Especiales del Instituto Español de Oceanografía 23: 153–160.

Crespi, V., 2002. Recent evolution of the fishing exploitation in the Thau lagoon, France. Fisheries Management and Ecology 9: 19–29.

De Casabianca, M. L., 1975. Production primaire et secondaire d'une lagune: Biguglia (Corse). Rapp Comm int Mer Medit 23: 63–64.

De Sostoa, A. & F. J. De Sostoa, 1985. The fish communities of the Ebro Delta (Northeast, Spain): a model of mediterranean estuarine ecosystem. In Yañez-Arancibia, A. (ed.), Fish Community Ecology in Estuaries and Coastal Lagoons: Towards and Ecosystem Integration, UNAM Press, México, 79–126.

Dowidar, N. M. & A. R. Abdel-Moati, 1983. Distribution of nutrient salts in lake Manzalah (Egypt). Rapp Comm int Mer Medit 28: 185–188.

Dowidar, N. M. & W. R. Hamza, 1983. Primary productivity and biomass of Lake Manzalah, Egypt. Rapp Comm int Mer Medit 28: 189–192.

Dumay, O., J. A. Tari, J. A. Tomasini & D. Mouillot, 2004. Functional groups of lagoon fish species in Languedoc Roussillon, southern France. Journal of Fish Biology 64: 970–983.

Fouda, M. M., M. Abboud-Abi Saab & M. A. Saleh, 1987. An extensive study of plankton in the Bardawil lagoon, Egypt. Lebanese Science Bulletin 3: 5–23.

Frisoni, G. -F., O. Guelorget, M. C. Ximenes & J. -P. Perthuisot, 1983. Etude ecologique de trois lagunes de la plaine orientale corse (Biguglia, Diana, Urbino):Expressions biologiques qualitatives et quantitatives du confinement. Journal de Recherche Oceanographique III bolletin 1: 57–80.

Gamito, S., J. Gilabert, C. Marcos & A. Pérez-Ruzafa, 2005. Effects of changing environmental conditions on lagoon ecology. In: Gönenç, I. E. & J. P. Wolflin (eds), Coastal Lagoons: Ecosystem Processes and Modeling for Sustainable Use and Development. CRC Press, Boca Ratón, 193–229.

González, J. A. & F. Conde, 1991. Estudio florístico, fenológico, autoecológico y fitogeográfico del macrofitobentos de la Mar Chica (Sebcha Buareg de Nador, Mediterráneo marroquí). Acta Botánica Malacitana 16: 63–80.

Gordo, L. S. & H. N. Cabral, 2001. The fish assemblage structure of a hydrologically altered coastal lagoon:the Óbidos lagoon (Portugal). Hydrobiologia 459: 125–133.

Gourret, M. P., 1897. Les étangs saumâtres du midi de la france et leurs pêcheries. Annales du Musée d'historie naturelle de Marseille -Zoologie V: 1–55.

Guelorget, O. & J. P. Perthuisot, 1983. Le domaine paralique. Expressions géologiques, biologiques et économiques du confinement. Travaux du laboratoire de géologie 16: 1–136.
Guelorget, O., G. F. Frisoni & J. P. Perthuisot, 1983. Zonation biologique des milieux lagunaires:definition d'une echelle de confinement dans le domaine paralique méditerranéen. Journal de recherche oceanographique 8: 15–35.
Halim, Y. & S. K. Guergues, 1981. Coastal lakes of the Nile Delta. Lake Manzalah. Unesco Technical papers in Marine Science 33: 135–172.
Hamon, P. Y., H. Tournier & P. Arnaud, 1979. Cycles annuels de quelques parametres physico-chimiques de L'Etang de Thau. Rapp Comm int Mer Medit 25/26.
Hervé, P. & J. Bruslé, 1980. L'étang de Salses-Leucate Écologie génénerale et ichtyofaune. Vie et Milieu 30: 275–283.
Hervé, P. & J. Bruslé, 1981. L'étang de Canet-Saint-Nazaire (P.O.) Écologie générale et ichthyofaune. Vie et Milieu 31: 17–25.
Izzo, G., A. Signorini & G Migliore, 2005. Progetto "Parchi in qualità" ovvero "applicazione pilota del Sistema di Gestione Ambientale nelle aree naturali protette"Analisi della qualità ecologica dei laghi costieri di Caprolace, Fogliano e Monaci. ENEA, Roma: 1–72.
Joyeux, J. C., J. A. Tomasini & J. L. Bouchereau, 1992. Modalités de la reproduction de *Gobius niger* (Teleostei, Gobiidae) dans une lagune méditerranéenne. Vie Milieu 42: 1–13.
Kapetsky, J. M., 1984. Coastal lagoon fisheries around the world: some perspectives on fishery yields, and other comparative fishery characteristcs. FAO Studies and Reviews 61(1): 97–139.
Kapetsky, J. M. & G. Lasserre, 1984. Management of coastal lagoon fisheries. General Fisheries Council for the Mediterranean Studies and Reviews 2 n° 61: 1–438.
Katselis, G., C. Koutsikopoulos, E. Dimitriou & Y. Rogdakis, 2003. Spatial patterns and temporal trends in the fishery landings of the Messolonghi-Etolikon lagoon system (western Greek coast). Scientia Marina 67: 501–511.
Kjerfve, B., 1994. Coastal Lagoon Processes. Elsevier Science Publishers, Amsterdam.
Kotsonias, G., 1984. The Messolonghi-Etolikon lagoon of Greece:Socio-economic and ecological interactions of cooperative and independent fishermen. FAO Studies and Reviews 61(1): 521–528.
Lasserre, G., 1989. Biologie halieutique des lagunes. Anné Biologique 28: 161–186.
Lemoalle, J. & G. Vidy, 1984. Conditions de milieu et peche dans la lagune hypersaline d'el Bibane (Tunisie). FAO Studies and Reviews 61(1): 175–195.
Leonardos, I. & A. Sinis, 1997. Fish mass mortality in the Etolikon lagoon, Greece:the role of local geology. Cybium 21: 201–206.
Lozano Cabo, F., 1953. Notas sobre una campaña de prospección pesquera en la mar chica de Melilla. Boletín del Instituto Español de Oceanografía 64: 1–37.
Mariani, S., 2001. Can spatial distribution of Ichthyofauna describe marine influence on Coastal lagoons? A central Mediterranean case study. Estuarine, Coastal and Shelf Science 52: 261–267.
McCullagh, P. & J. A. Nelder, 1989. Generalized Linear Models, 2nd edn. Chapman and Hall, New York.
Menéndez, M., J. Herrera & F. A. Comín, 2002. Effect of nitrogen and phosphorus supply and growth, chlorophyll content and tissue composition of the macroalga *Chaetomorpha linum* (O.F.Müll.) Kütz in a mediterranean coastal lagoon. Scientia Marina 66 (4): 355–364.
Miller, J. M., L. J. Pietrafesa & N. P. Smith, 1990. Principles of Hydraulic Management of Coastal Lagoons for Aquaculture and Fisheries. FAO Fisheries Technical Paper 314. FAO, Rome.
Mistri, M., E. A. Fano, F. Ghion & R. Rossi, 2002. Disturbance and community pattern of polichaetes inhabiting Valle Magnavacca (Valli di Comacchio, Northern Adriatic sea, Italy). Marine Ecology 23: 31–49.
Nixon, S. W., 1982. Nutrient dynamics, primary production and fisheries yields of lagoons. Oceanologica Acta, Special edition, 357–371.
Paris, J. & J. P. Quignard, 1971. La faune ichthyologique des étangs languedociens de Sète a Carnon (Écologie, Éthologie). Vie et Milieu 22: 301–327.
Pejá, N., A. Vaso, A. Miho, N. Rakaj & A. J. Crivelli, 1996. Characteristics of Albanian lagoons and their fisheries. Fisheries Research 27: 215–225.
Pérez-Ruzafa, A., 1989. Estudio ecológico y bionómico de los poblamientos bentónicos del Mar Menor (Murcia, SE de España). Unpublished PhD. Universidad de Murcia.
Pérez-Ruzafa, A. & C. Marcos, 1992. Colonization rates and dispersal as essential parameters in the confinement theory to explain the structure and horizontal zonation of lagoon benthic assemblages. Rapp Comm int Mer Medit 33: 100.
Pérez-Ruzafa, A. & C. Marcos, 1993. La teoría del confinamiento como modelo para explicar la estructura y zonación horizontal de las comunidades bentónicas en las lagunas costeras. Publicaciones Especiales del Instituto Español de Oceanografía 11: 347–358.
Pérez-Ruzafa, A., J. Gilabert, J. M. Gutiérrez, A. I. Fernández, C. Marcos & S. Sabah, 2002. Evidence of a planktonic food web response to changes in nutrient input dynamics in the Mar Menor coastal lagoon, Spain. Hydrobiologia 475/476: 359–369.
Pérez-Ruzafa, A., J. I. Quispe-Becerra, J. A. García-Charton & C. Marcos, 2004. Composition, structure and distribution of the ichthyoplankton in a Mediterranean coastal lagoon. Journal of Fish Biology 64: 202–218.
Pérez-Ruzafa, A., A. I. Fernández, C. Marcos, J. Gilabert, J. I. Quispe-Becerra & J. A. García-Charton, 2005. Spatial and temporal variations of hydrological conditions, nutrients and chlorophyll *a* in a Mediterranean coastal lagoon: (Mar Menor, Spain). Hydrobiologia 550: 11–27.
Pérez-Ruzafa, A., J. A. García-Charton, E. Barcala & C. Marcos, 2006. Changes in benthic fish assemblages as

a consequence of coastal works in a coastal lagoon: the Mar Menor (Spain, Western Mediterranean). Marine Pollution Bulletin 53: 107–120.
Petit, G., 1953. Introduction à l'étude écologique des étangs méditerranéens. Vie Milieu 4: 569–604.
Poizat, G., E. Rosecchi, P. Chauvelon, P. Contournet & A. J. Crivelli, 2004. Long-term fish and macro-crustacean community variation in a Mediterranean lagoon. Estuarine, Coastal and Shelf Science 59: 615–624.
Por, F. D., 1972. Hidrobiological notes on the high-salinity waters of the Sinai Peninsula. Marine Biology 14: 111–119.
Por, F. D., 1980. A classification of hypersaline waters, based on trophic criteria. PSZN I:Marine Ecology 1: 121–131.
Por, F. D. & A. Ben-Tuvia, 1981. The Bardawil lagoon (Sirbonian lagoon) of north Sinai—a summing up. Rapp Comm int Mer Medit 27: 101–107.
Quignard, J. P., 1984. Les caracteristiques biologiques et environnementales des lagunes en tant que base biologique de l'amenagement des pecheries. FAO Studies and Reviews 61(1): 3–38.
Raimbault, R., 1972. Données hydrologiques sur les étangs corses de Diana et d'Urbino (1969–1970). Rapp Comm int Mer Medit 21: 117.
Riouall, R., 1977. Évolution de la flore benthique des étangs de Berre et Vaine (Bouches-du-Rhone) de 1971 a 1975. Bull Mus Hist Nat Marseille 37: 141–147.
Rosecchi, E., F. Thomas & J. Crivelli, 2001. Can life-history traits predict the fate of introduced species? A case study on two cyprinid fish in southern France. Freshwater Biology 46: 845–853.
Rossi, R., V. U. Ceccherelli, I. Ferrari, V. Gaiani, L. Carnacina & G. Colombo, 1984. Management of aquaculture and fishing in the Scardovari fishery (Po River Delta, Italy). FAO Studies and Reviews 61(2): 441–459.
Rowntree, J., 1984. Fisheries management in the northern Nile Delta lakes of Egypt: the case of Hosha. FAO Studies and Reviews 61(2): 541–556.
Sioku-Frangou, I. & O. Gotsis-Skretas, 1985. Aspect du plancton de la lagune de Messolongui (Hellas) en juin 1983. Rapp Comm int Mer Medit 29: 171–173.
Soria, J. M., M. R. Miracle & E. Vicente, 2002. Relations between physico-chemical and biological variables in aquatic ecosystems of the Albufera Natural Park (Valencia, Spain). Verh Internat Verein Limnol 28: 564–568.
Suzuki, M. S., R. O. Figueiredo, S. C. Castro, C. F. Silva, E. A. Pereira, J. A. Silva & G. T. Aragon, 2002. Sand bar opening in a coastal lagoon (Iquipari) in the northern region of Rio de Janeiro state:hydrological and hydrochemical changes. Brazilian Journal of Biology 62: 51–62.
ter Braak, C. J. F., 1990. Update Notes: CANOCO v3.10. Agricultural Mathematics Group, Wageningen.
Tilman, D., 1982. Resource, Competition and Community Structure. Princenton University Press, Princenton.
Tilman, D., 1999. The ecological consequences of changes in biodiversity: a search for general principles. Ecology 80: 1455–1474.
Tournier, H., P. Y. Hamon & S. Landrein, 1983. Conditions de milleu moyennes dans L'Étang de Thau établies sur les observations réalisées de 1974 à 1980. Rapp Comm int Mer Medit 28: 195–200.
UNESCO, 1980. Coastal lagoons survey. UNESCO Technical Papers in Marine Science, 31: 7.
UNESCO, 1981. Coastal lagoons research, present and future. UNESCO Technical Papers in Marine Science 32: 51–79.
Vandenm Bossche, J. P. & G. M. Bernacsek, 1991. Source book for the inland fishery resources of Africa. CIFA Technical Paper 18. FAO Roma, 1–219.
Villena, M. J. & S. Romo, 2003. Phytoplankton changes in a shallow Mediterranean lake (Albufera of Valencia, Spain) after sewage diversion. Hydrobiologia 506–509: 281–287.
Yáñez-Arancibia, A. & R. S. Nugent, 1977. El papel ecológico de los peces en estuarios y lagunas costeras. An Centro Cienc del Mar y Limnol Univ Nal Autón México 4: 107–114.

Hydrobiologia (2007) 577:127–139
DOI 10.1007/s10750-006-0422-7

LAGOONS AND COASTAL WETLANDS

Species–area patterns of benthic macro-invertebrates in Italian lagoons

Letizia Sabetta · Enrico Barbone · Agnese Giardino · Nicola Galuppo · Alberto Basset

Abstract The selection of adequate descriptors of the ecological status in aquatic ecosystem is a major requirement for the implementation of monitoring tools. It requires an analysis of the stress-independent sources of variation of potential descriptors, which need to be taken into account in the definition of ecosystem Typology and Classification. Here, we investigate at what extent the surface area of lagoons accounted for species richness of communities and body size abundance components. To this aim, the species–area and body size–area patterns of benthic macro-invertebrates were investigated in Italian lagoons. The analysis was based on a literature survey carried out considering a 30-year period from 1975 to 2004. Overall, 168 papers were selected, reporting taxonomic lists of benthic macro-invertebrates for 26 Italian lagoons, whose surface area ranged from few hectares to some hundreds of square kilometres. The analysis of published taxonomic lists recorded 1,055 taxa, belonging to 13 phyla, 106 orders and 351 families. Measures of standard body size for each of the 1,055 taxa were also obtained from published data as standard individual body length. Significant species–area relationships were observed at each level of taxonomic resolution considered, within dominant phyla (i.e. Arthropoda and Mollusca). Slopes of the power regressions were in the range of 0.11–0.24, increasing consistently with taxonomic resolution. Significant relationships were also observed by comparing the upper limit of the body size spectra to the surface area of the considered lagoons. Maximum body size–area relationships were described by power regressions with slopes in the range of 0.10–0.28. The observed species–area relationships underline the importance of physiographic characteristics of transitional water ecosystems in the identification of transitional water types, as required by the WFD. Moreover, the scaling of average taxonomic richness with lagoon surface, could also represent a standardisation tool for classifying the ecological status of transitional ecosystems. Finally, the patterns of body size area relationships would also have an important application to the field of monitoring transitional ecosystem health.

Keywords Italian lagoons · Benthic macro-invertebrates · Island biogeography · Species–area relationships · Body size–area relationships · Water Framework Directive

Guest editors: P. Viaroli, P. Lasserre & P. Campostrini
Lagoons and Coastal Wetlands in the Global Change Context: Impacts and Management Issues

L. Sabetta (✉) · E. Barbone · A. Giardino · N. Galuppo · A. Basset
Department of Biological and Environmental Sciences and Technologies, University of Lecce, S.P. Lecce-Monteroni, 73100 Lecce, Italy
e-mail: letizia.sabetta@unile.it

Introduction

The tendency for species richness to increase with area (the "species–area relationship") is one of the most robust empirical generalisations in ecology (MacArthur & Wilson, 1967; Holt et al., 1999). Conventionally, species–area relationships have an *immigration component*, which combines a "sample effect" on the regional species pool (Connor & McCoy, 1979) with meta-community level processes of patch dynamics and species sorting (Leibold et al., 2004), and a *community level component*, based on assembly rules, niche partitioning and niche breadth plasticity (MacArthur, 1972). Habitat productivity and heterogeneity are the two major factors determining the community level component. Larger areas have higher overall productivity and habitat heterogeneity. Species richness tends to be directly related to habitat heterogeneity (Williams, 1964) and to energy availability (e.g. Wright, 1983; Srivastava & Lawton, 1998; Gaston 2000), which affects population densities and the resulting risk of extinction (Simberloff & Abeie, 1976; Schoener & Spiller, 1992) and increases the occurrence of resources enabling the persistence of viable populations (Bonn et al., 2004). However, species richness is also strongly affected by external perturbations and pollution, which can have a major influence on the species–area relationships where relevant anthropogenic pressures occur (e.g. Rosenberg & Resch, 1993).

In the theory of island biogeography the term "islands" can be used to refer not only to pieces of land surrounded by water, but to each 'island' system where similar habitat types are separated in space by relatively unfavourable habitats.

Transitional waters, which can be defined as *"bodies of surface water in the vicinity of river mouths which are partly saline in character as a result of their proximity to coastal waters but which are substantially influenced by freshwater flows"* (European Water Framework Directive, EU-WFD 60/2000), are habitat islands in the coastal landscape. Transitional waters include a number of different habitat types: i.e. estuaries, deltas, lagoons, coastal lakes and ponds, brackish wetlands and salt-marshes. Enclosed bays receiving freshwater inflows can also be considered transitional waters.

Transitional waters are a well-defined type of aquatic ecosystem, whose structural abiotic characteristics arise from their origin as ecotones between terrestrial, freshwater and marine ecosystems (Carrada et al., 1988; Basset & Abbiati, 2004). On the other hand, taxa exclusively occurring in transitional waters are rare (Guelorget & Perthuisot, 1983), since transitional water communities are strongly influenced by freshwater and marine communities. Both plant and animal taxa immigrate from the freshwater or marine environments; thus, the lack of any physical barrier between the transitional ecosystems and their input and output environment poses the question as to whether transitional waters can be actually considered habitat islands for both animal and plant guilds or if they can simply be considered relatively enclosed marine habitats or coastal freshwater water bodies depending on their hydrology and tidal range.

Species–area relationships have been already addressed in freshwater ecosystems (e.g. lakes: Elmberg et al., 1994; Griffiths, 1997; Amarasinghe & Welcomme, 2002), while they remain largely unexplored in transitional waters. For these environments, the analysis of species–area relationships can have important applications, since species richness is a commonly used descriptor of the health of these ecosystems. The assessment of the ecological status of transitional waters, through the evaluation of different components of the ecosystem, is a key requirement for the WFD for each water body. This goal has to be reached by a series of sequential steps, starting with the definition of a water body typology, which then leads to the classification of the water body's ecological status. WFD requires the definition of an aquatic ecosystem typology in order to minimise intra-type variability of the ecological status descriptors. In this context, analysis of expected species/area relationships could represent a valuable tool in the classification of transitional ecosystem ecological status.

Here, the species–area relationship of macro-invertebrate communities in transitional aquatic ecosystems in Italian lagoons was investigated. The main aims of the study were: (i) to describe the structure of benthic macro-invertebrate guilds in the Italian lagoons; (ii) to analyse the species–area relationships in benthic macro-invertebrate guilds as a function of the size of the available dataset and the level of taxonomic resolution; and (iii) to evaluate the occurrence of body size–area relationships.

Materials and methods

Data sources

This paper is based on a data mining activity on the taxonomic composition of benthic macro-invertebrate guilds in Italian lagoons. The data used in the present study were gleaned from published sources, including international journals retrieved from electronic databases (i.e. *ASFA*, *WebSpirs*, *Biological Abstracts, Current Contents* and *Zoological Records*), Italian journals covering the fields of ecology and zoology (i.e. *Accademia Peloritana, Il naturalista siciliano, Oebalia, Rendiconti del Seminario della Facotà di Scienze dell'Università di Cagliari, Thalassia Salentina*) and proceedings of Italian scientific societies covering the field of aquatic ecology and zoology (i.e. *Associazione Italiana di Oceanografia e Limnologia, Società Italiana di Biologia Marina, Società Italiana di Ecologia, Società Sarda di Scienze Naturali*). The search was limited to papers published in the last 30 years.

The electronic search was performed through a three-way factorial combination of the following groups of keywords: (a) Italy, Italian; (b) lagoon, coastal lake, coastal pond, salt-marsh, saltern, brackish; (c) benthos, macrobenthos, macro-invertebrate, benthic invertebrates, benthic fauna. The search on Italian journals and proceedings was performed through a check on the index of content of all the issues published in the considered period and, when required, of the abstracts. Three unpublished reports to which one of the authors directly contributed were also taken into consideration.

The dataset

In total the search produced 205 papers containing taxonomic lists of benthic macro-invertebrates in Italian lagoons; most papers focused on a single lagoon but comparative studies on different lagoons were also collected in this way.

The 205 papers were screened according to three main criteria: (a) taxonomic resolution (most taxa classified down to the genus or species level); (b) taxonomic completeness (the taxonomic list of a lagoon, resulting from one or more papers, cannot be limited to a few selected phyla); (c) sampling design (samplings carried out at a seasonal or higher resolution). According to these criteria 168 papers were selected for the analysis, referring to 26 lagoon ecosystems. A list of the considered ecosystems, reporting their surface area and the number of published papers collected is presented in Fig. 1. In the dataset obtained by applying these criteria, the number of articles referring to each of the 26 transitional aquatic ecosystems ranges from 1 (i.e. Massaciuccoli coastal lake, Torre Guaceto brackish wetland, Marsala saltern and Piallassa Baiona lagoon) to 22 (i.e. Venice lagoon).

A more selective screening of the 205 collected papers was also performed with the addition of a fourth criterion to the three quoted above, i.e., a minimum number of five papers per transitional aquatic ecosystem (>5 papers). This further selection reduced the number of selected papers and, accordingly, the number of transitional water ecosystems taken into consideration; moreover, the inter-system heterogeneity in the amount of information collected was reduced. The dataset obtained according to the four selective criteria includes 16 transitional aquatic ecosystems (reduced dataset hereafter) out of the 26 selected by applying only the first three criteria (complete dataset).

For each lagoon 15 physiographic and hydrological parameters were collected from the published papers as well as from published satellite images and from the "Tide tables of Italy" (Istituto Idrografico della Marina, 1999). The parameters were surface area, perimeter and sinuosity of the water body, maximum axis and minimum axis, measures of length and width of its outlet(s) and their ratio (length/width), measures of mean depth, maximum and minimum tide, tidal range, maximum and minimum salinity and salinity range.

ID	Lagoon			Surface	N°
	Name	Region	Province	(km^2)	papers
1	Grado Marano Lagoon	Friuli - Venezia Giulia	GO	160	6
2	Venezia Lagoon	Veneto	VE	549	22
3	Canarin Lagoon	Veneto	RO	8.9	8
4	Scardovari Lagoon	Veneto	RO	38	5
5	Lake Nazioni	Emilia Romagna	FE	4	2
6	Goro Lagoon	Emilia Romagna	FE	32	6
7	Valli di Comacchio	Emilia Romagna	FE	115	11
8	Pialassa Baiona Lagoon	Emilia Romagna	RA	11.8	1
9	Lake Massaciuccoli	Tuscany	LU	6.9	1
10	Orbetello Lagoon	Tuscany	GR	27	12
11	Lake Lesina	Apulia	FG	51	7
12	Lake Varano	Apulia	FG	60.5	3
13	Lake Fogliano	Latium	LT	4	8
14	Lake Monaci	Latium	LT	0.95	8
15	Lake Caprolace	Latium	LT	2.3	8
16	Lake Sabaudia	Latium	LT	3.7	8
17	Lake Fondi	Latium	LT	3.9	7
18	Lake Lungo	Latium	LT	0.47	7
19	Lake Fusaro	Campania	NA	1	8
20	Torre Guaceto Brackish Area	Apulia	BR	1.2	1
21	Acquatina Lagoon	Apulia	LE	0.45	5
22	Lake Alimini Grande	Apulia	LE	1.3	7
23	Oliveri Tindari Brackish Area	Sicily	ME	0.19	2
24	Saltern of Marsala	Sicily	TP	2.37	1
25	Marsala Lagoon	Sicily	TP	24	9
26	Rada di Augusta	Sicily	ME	23.5	4

Fig. 1 Geographical distribution of selected transitional aquatic ecosystems among regions and provinces in Italy. A list of lagoons with their surface area and number of papers retrieved is reported

Standard body length of benthic macro-invertebrate taxa recorded in the 26 lagoons was also obtained from literature sources: i.e. classification keys and books on zoology of invertebrates. Overall, standard body length was retrieved from the literature for 420 out of the 1,055 taxa. Therefore all the analyses of body size–area relationships are based on a sub-sample of 420 taxa for which body size information was available.

Study sites

The 26 Italian coastal lagoons for which published species lists of benthic macro-invertebrate taxa were collected are distributed among 14 provinces belonging to eight regions (Fig. 1), with six coastal lagoons in the region of Lazio, five in the region of Puglia, four in the regions of Sicilia and Emilia Romagna, three in the regions of Veneto and Toscana and one in the region of Friuli Venezia Giulia.

In total, the 26 selected lagoons cover a coastal area of 1,133.43 km^2, which represents 69.6% of the overall surface of Italian transitional waters. The restricted set of 16 lagoons cover a coastal area of 984.50 km^2, which represents 60.5% of the overall surface of Italian transitional waters.

Data analysis

The data were organised into two matrices, a presence/absence matrix with benthic macro-invertebrate data was organised into 26 columns (i.e. the selected lagoons) and 1,055 rows (i.e. macro-invertebrate taxa); and a rectangular matrix with 26 rows (i.e. the selected lagoons) and 17 columns, with two columns describing the benthic macro-invertebrate guilds (i.e. taxonomic richness and standard body length of the largest species), nine columns describing physiographic characteristics of the water bodies (i.e. surface area, perimeter, sinuosity, length of min. axis, length of max. axis, outlet length, outlet width, outlet length/width ratio, average water depth), and three columns describing the tidal regime (i.e. minimum and maximum tides and tidal range) and the columns describing salinity regime (i.e. minimum, maximum and range of salinity).

The taxonomic composition similarity between lagoon pairs was measured using the Jaccard similarity index, as follows:

$$S_J = [a/(a + b + c)] \times 100,$$

where S_J, Jaccard similarity coefficient; a, number of taxa shared by two lagoons; b, number of taxa unique to the first lagoon; and *c*, number of taxa unique to the second lagoon.

PCA was used as an explorative analysis to evaluate the role of the abiotic factors in the structural heterogeneity of the selected sample of Italian lagoons.

The species area–relationships in the selected sample of Italian lagoons were analysed utilising the Arrenhius power function (Arrenhius, 1921) as a descriptive model for species accumulation patterns:

$$S = cA^{Z},$$

or

$$\log S = \log C + z \log A,$$

where A is the surface area of the lagoon, S is taxonomic richness expressed as total number of species found, and c and z are regression coefficients. Specifically, z indicates the rate of species number increasing with area and c is a fitting coefficient, or estimated number of species per "unit" area.

We analysed the species area relationships in the two groups of lagoons selected according to the different criteria reported above and for each group of lagoons we analysed the species–area relationships in relation to the taxonomic resolution (i.e. at the level of taxa, family, order) and for each of the most important macro-invertebrate phyla in transitional aquatic ecosystem (i.e. Annelida, Arthropoda, Mollusca).

The Arrenhius power function was also used to analyse the body size–area relationships, where the taxonomic richness (S) was substituted with the standard length of the largest taxon within each macro-invertebrate community.

Multiple regression was used to evaluate the relative influence and the cumulative importance of surface area and other abiotic structural factors to macro-invertebrate taxonomic richness in the studied Italian lagoons.

Results

Features of selected study sites

The sample of 26 lagoons selected for this study includes some of the largest Italian lagoons, occurring along the Northern Adriatic coast, as well as some Southern small lagoons, lagoons with a tidal range of more than 1 m, close to the highest tidal range of the Mediterranean region, and lagoons with only few centimetres of tidal range. Surface area ranged from 0.19 km^2 [i.e. the lagoon system of Oliveri-Tindari (ME)] to 549 km^2 [i.e. Venice Lagoon (VE)] (Fig. 1).

Overall, the PCA on the structural abiotic factors considered in the study accounted for 77% of the difference among lagoons with four component axes; the first three axes accounted for the physiographic characteristics of the water body, water salinity and the physiographic characteristics of the water body outlet (Table 1).

Macro-invertebrate guild composition

The taxonomic list of benthic macro-invertebrates in the considered sample of Italian lagoons comprised 1,055 benthic macro-invertebrate taxa belonging to 13 phyla (Mollusca, Annelida, Arthropoda, Briozoa, Cnidaria, Porifera, Tunicata, Echinodermata, Nematoda, Platyhelmintha, Entoprocta, Nemertea, Phoronida), 106 orders and 351 families. Mollusca, Annelida and Arthropoda

Table 1 PCA of the structural abiotic factors considered in the study. The contribution of each variable to the variance of the extracted components and the cumulative percentage of variance accounted for by the four extracted components are shown

	Components			
	1	2	3	4
Surface area	0.91	0.28	0.09	–0.09
Perimeter	0.92	0.25	0.14	–0.15
Sinuosity	–0.05	–0.08	0.44	–0.42
Length of max. axis	0.93	0.20	0.11	–0.12
Length of min. axis	0.73	0.15	–0.02	–0.10
Outlet length	–0.27	–0.30	0.72	0.08
Outlet width	0.85	0.36	0.09	–0.09
Outlet length/width ratio	–0.18	–0.24	0.59	0.56
Average water depth	–0.23	0.44	0.49	–0.23
Maximum tide	0.82	–0.31	0.09	0.37
Minimum tide	0.52	–0.28	0.05	0.47
Tidal range	0.80	–0.27	0.10	0.24
Minimum salinity	–0.29	0.86	0.05	0.38
Maximum salinity	–0.13	0.57	–0.60	0.40
Salinity range	0.28	–0.67	–0.62	–0.16
Cumulative explained variance (%)	*38.15*	*54.42*	*68.51*	*77.55*

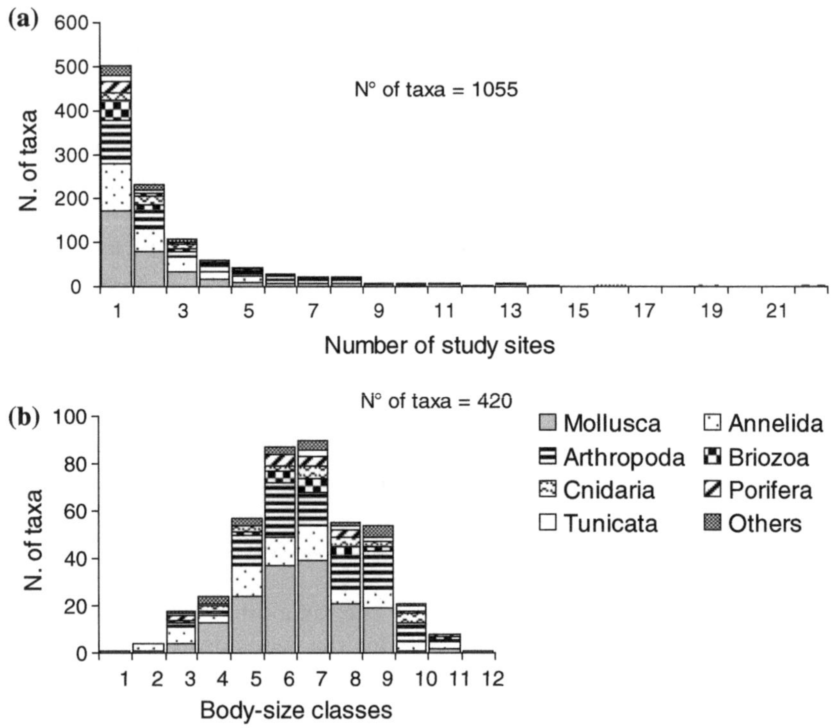

Fig. 2 (**a**) Distribution of taxa among the 26 transitional aquatic ecosystems considered in this study: number of taxa found in given numbers of lagoons. (**b**) Absolute frequency distribution of taxa among 12 logarithmic classes of standard body length derived from published values. Body size class 1 = 0.9 mm; 2 = 1.7 mm; 3 = 3 mm; 4 = 6 mm; 5 = 11 mm; 6 = 20 mm; 7 = 36 mm; 8 = 66 mm; 9 = 120 mm; 10 = 219 mm; 11 = 398 mm; 12 = 724 mm

were the dominant phyla among the 13 identified, collectively accounting for 77% of the 1,055 taxa.

Most taxa had very restricted spatial distributions; more than 65% of the macro-invertebrate taxa were observed only in a single ecosystem, and 21% of taxa were observed only in two coastal lagoons (Fig. 2a). The most widely distributed species, mollusca *Cerastoderma glaucum* (Poiret, 1789), was observed at 21 out of the 26 considered ecosystems. At a local scale, taxonomic richness ranged from 18 to 356 macro-invertebrate taxa per ecosystem (Table 2). The taxonomic composition of macro-invertebrate guilds varied greatly among ecosystems; indeed, the average inter-lagoon similarity in taxonomic composition was as low as 0.086 ± 0.004.

Standard body length of benthic macro-invertebrate varied greatly within guilds. Overall variation encompassed three orders of magnitude, with the macro-invertebrate taxa distributed into 12 logarithmic body length classes (Fig. 2b); most taxa (i.e. 177 taxa) had an intermediate body length, between 3.5 mm and 5.0 mm.

Species–area and body size–area distributions

The number of taxonomic units increased with the surface area of the Italian lagoons at each of the three levels of taxonomic resolution considered and in both groups of lagoons selected (Figs. 3a, 4a). Both for the 26 and for the 16-lagoon set, the slope of the species–area regressions decreased with taxonomic resolution, ranging from 0.19 (resolution at the taxa level) to 0.12 (resolution at the order level) in the group of 26 lagoons, and from 0.20 (resolution at the taxa level) to 0.10 (resolution at the order level) in the group of 16 lagoons.

The number of taxonomic units increased with the surface area of the Italian lagoons also within two out of the three most representative phyla:

Table 2 Number of benthic macro-invertebrate taxa recognised for each lagoon

ID	Lagoon	No. of taxa
1	Grado Marano Lagoon	154
2	Venezia Lagoon	356
3	Canarin Lagoon	74
4	Scardovari Lagoon	60
5	Lake Nazioni	20
6	Goro Lagoon	154
7	Valli di Comacchio	183
8	Pialassa Baiona Lagoon	27
9	Lake Massaciuccoli	18
10	Orbetello Lagoon	210
11	Lake Lesina	117
12	Lake Varano	65
13	Lake Fogliano	85
14	Lake Monaci	46
15	Lake Caprolace	135
16	Lake Sabaudia	42
17	Lake Fondi	42
18	Lake Lungo	68
19	Lake Fusaro	154
20	Torre Guaceto Brackish Area	35
21	Acquatina Lagoon	71
22	Lake Alimini Grande	129
23	Oliveri Tindari Brackish Area	61
24	Saltern of Marsala	24
25	Marsala Lagoon	256
26	Rada di Augusta	149

i.e. Arthropoda and Mollusca (Figs. 3b, 4b). The slope of the species–area regressions for these phyla varied narrowly from 0.24 to 0.21 in both groups of lagoons. On the other hand, significant species–area relationships were not observed within the Annelida, which represents the third most representative phylum.

The significance level of the species–area regressions was increased by extending the regression model to a multiple regression including the outlet length as an independent variable (Fig. 5).

Body size–area regressions were also observed both in the 26 and the 16-lagoon data set. In the 16-lagoon data set, maximum length (the standard length of the largest observed species in each lagoon) significantly increased with the lagoon surface area in the sub-samples of Arthropoda, Bivalvia and other taxa (including all the 10 less representative phyla) (Fig. 6); Significant body size–area regressions were also observed in the 26-lagoon data set for the sub-samples of Arthropoda and other taxa (Fig. 7). The slopes of the body size area regressions were in a narrow range of variation between 0.19 and 0.24 in the 16-lagoon data set.

Discussion

The large scale of macro-ecological patterns does not allow simple direct experimental testing and this is generally based on the analysis of literature surveys. Here, using the literature survey

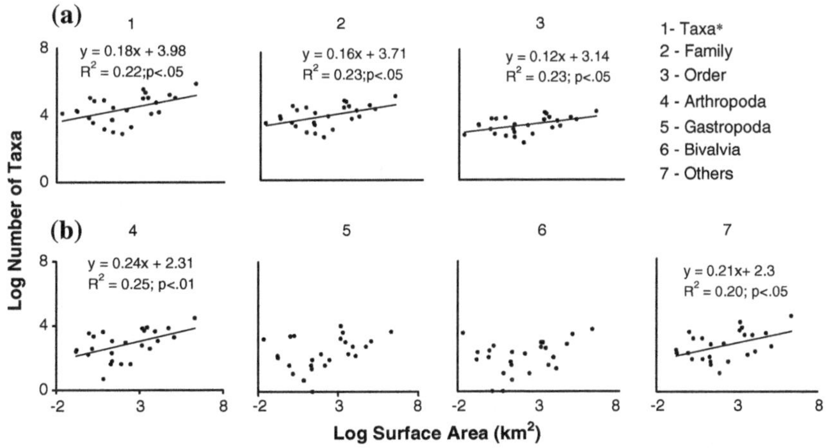

Fig. 3 Relationships between number of taxa and surface area of the Italian lagoons at each of the three levels of taxonomic resolution considered (**a**) and within three dominant phyla (**b**) for the total data set (26 lagoons). *Taxa means that, for each taxonomic units retrieved, the higher taxonomic resolution was considered

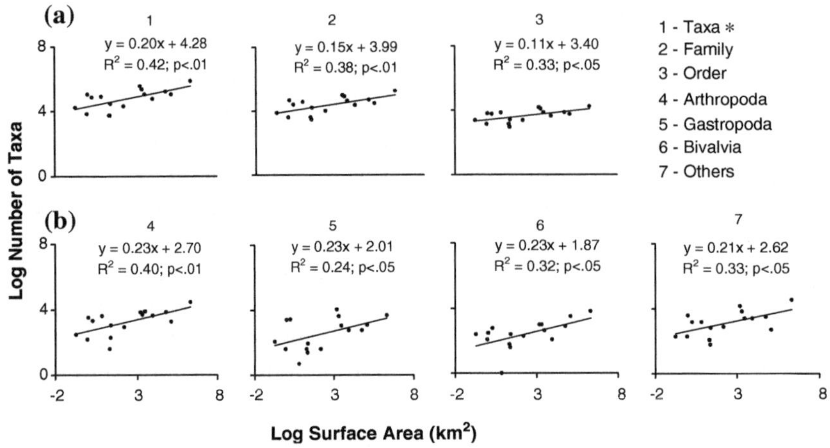

Fig. 4 Relationships between number of taxa and surface area of the Italian lagoons at each of the three levels of taxonomic resolution considered (**a**) and within three dominant phyla (**b**) for the total data set (16 lagoons). *Taxa means that, for each taxonomic units, retrieved the higher taxonomic resolution was considered

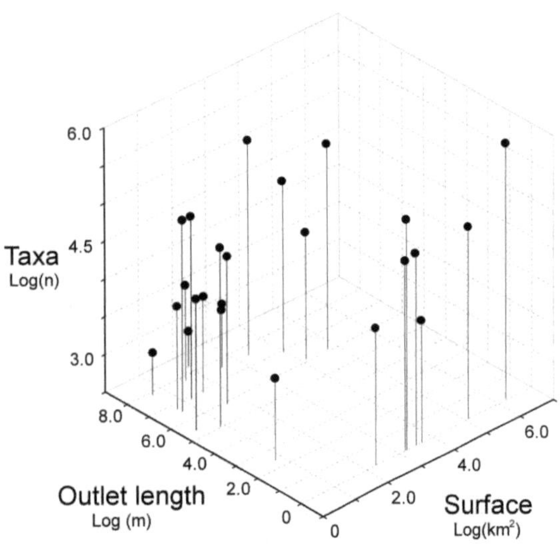

Fig. 5 Multiple regression including the outlet length and surface area as independent variables and number of macro-invertebrate taxa as a dependent variable

approach with Italian lagoons, two major macro-ecological patterns of macro-invertebrate distribution in lagoon ecosystems were emphasised:

1. The taxonomic heterogeneity of macro-invertebrate guild compositions of different lagoons; and
2. The consistency of species–area and body size–area distributions of benthic macro-invertebrates.

The first macro-ecological pattern arose from the analysis of the macro-invertebrate distributions in the 26 considered lagoons, which were likely to be a representative sample of Italian lagoons, accounting for 14.8% of lagoon numbers and 69.6% of lagoon surface in Italy (Basset & Sabetta, 2005). In fact, 86% of the macro-invertebrate taxa were observed to occur in less than three lagoons and low similarity was observed also between neighbouring lagoons of similar surface area which would be expected to share a common potential species pool, such as Lesina and Varano in the Puglia region ($S_J = 0.174$; Jaccard similarity index) or Caprolace and Monaci in the Lazio region ($S_J = 0.154$; Jaccard similarity index).

In principle, the diversity of macro-invertebrate guilds could result from stochastic ecological and evolutionary processes on the local (e.g. community level) and regional (e.g. meta-community level) scale (Hubbell, 2001), or from dispersal and recruitment limitations on macro-invertebrate taxa (Hurtt & Pacala, 1995), or from the structural heterogeneity of the transitional waters (Basset & Abbiati, 2004) and the consequent selection of macro-invertebrate taxa according to their functional traits and niche requirements (MacArthur, 1970).

A neutral theory explanation of macro-invertebrate guild heterogeneity in Italian lagoons did not

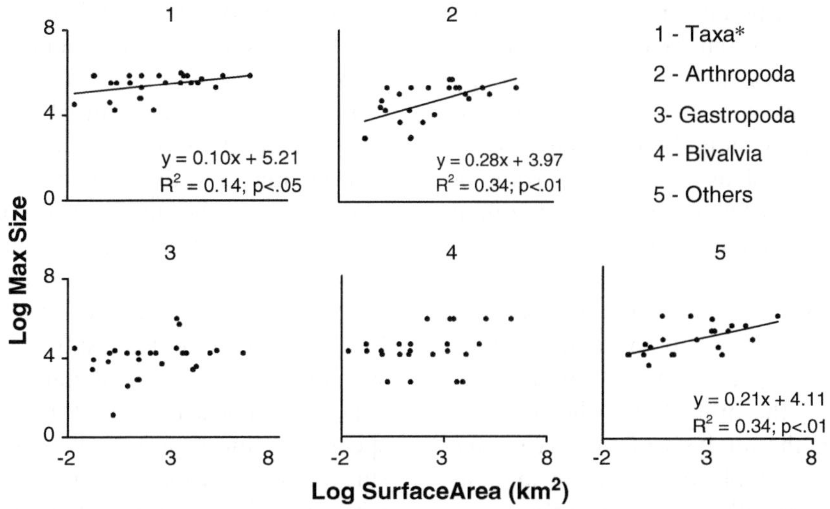

Fig. 6 Relationships between maximum body-size (standard length of the largest observed species in each lagoon) and surface area of the Italian lagoons at the highest taxonomic resolution considered, within three dominant phyla separately and the other phyla together for the total data set (26 lagoons). *Taxa means that, for each taxonomic units, retrieved the higher taxonomic resolution was considered

seem to be supported by the data. The fact that heterogeneity of guild taxonomic composition among lagoons was statistically different than that expected by chance (expected similarity, $S_J = 0.041 \pm 0.002$; observed similarity, $S_J = 0.086 \pm 0.004$; *t*-Student test, $t = 11.5$; d.f. = 324; $P < 0.01$) and that the observed similarity among guild taxonomic compositions showed consistent non-random geographical clustering (Sabetta et al., 2005) supported this point. The patterns of taxonomic composition similarity among lagoons did not appear to be accounted for by potential limitations on the dispersal and recruitment of macro-invertebrate taxa. Indeed, even though neighbouring lagoons, which would be expected to share a common hydrology and common

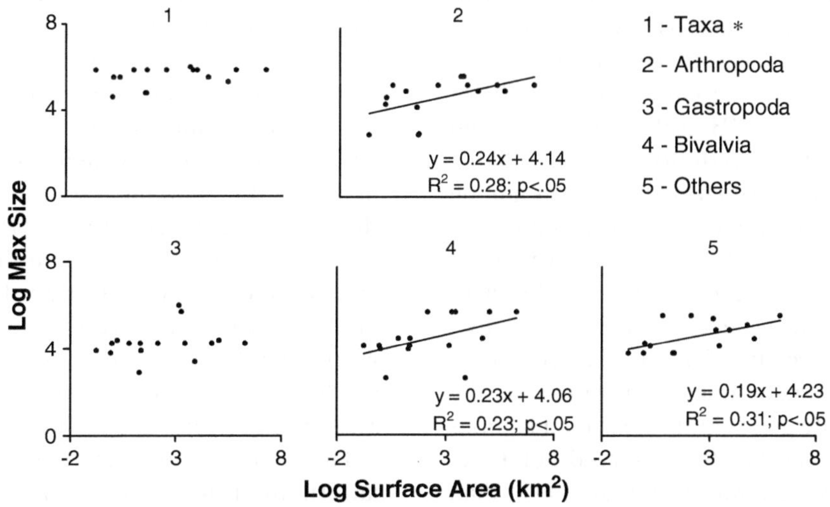

Fig. 7 Relationships between maximum body-size (standard length of the largest observed species in each lagoon) and surface area of the Italian lagoons at the highest taxonomic resolution considered, within three dominant phyla separately and the other phyla together for the total data set (16 lagoons). *Taxa means, that for each taxonomic units, retrieved the higher taxonomic resolution was considered

recruitment from marine and freshwater environments, showed more similar taxonomic composition than lagoons which were geographically distant, this similarity was in any case very low, even when comparing neighbouring lagoons of similar surface area, such as Lesina and Varano or Caprolace and Monaci. Dispersal and recruitment limitation could explain the meta-community level heterogeneity and could have a role in the geographical clustering of Italian lagoons (Sabetta et al., 2005), but they did not seem to account for the variations observed at a smaller spatial geographical scale. On the other hand, the observations arising from the analysis of the taxonomic composition of the sample of 26 Italian lagoons supported the third hypothesis proposed above, i.e. the niche theory hypothesis. Within and among habitats, structural heterogeneity is a common feature of lagoons and coastal lakes, which are ecotones between terrestrial, freshwater and marine ecosystems (e.g. Basset & Abbiati, 2004), whose relative influence varies through time and determines characteristic scales of temporal variability inside transitional aquatic ecosystems (Comin et al., 2004). Various classifications have been proposed to account for the heterogeneity of transitional Mediterranean waters, based on geomorphology (Kjerfve, 1994), salinity (Battaglia, 1959) and degree of confinement (Guelorget & Perthuisot, 1983; Guelorget et al., 1983). Physiographic characteristics, such as surface area and outlet length, have also been recently proposed as major factors affecting lagoon typology both for the Mediterranean area (Basset et al., 2006) and for Polynesian area (Andrefouet et al., 2001). Physiographic and hydrological differences among lagoons set different environmental niche conditions within which there is ground for niche partitioning among species. Niche differences among species with respect to the physical environment could account for the heterogeneity observed in the taxonomic composition of the macro-invertebrate guilds in the 26 Italian lagoons considered. The fact that the observed heterogeneity, though very high, was in any case lower than would be expected from a random process of taxonomic composition of macro-invertebrate guilds supported this explanation, suggesting that the distribution of the most widely distributed taxa was related to a deterministic selection of habitat patches, or sets of environmental niche conditions, within the lagoon ecosystems.

From an applied point of view, the heterogeneity of guild taxonomic composition in Italian lagoons makes it difficult to use taxonomy-based indicators in the monitoring of the ecological status of Italian lagoons; because they assume presence or relative abundance of specific taxa.

The analysis of macro-invertebrate guilds of Italian lagoons and coastal lakes emphasised a second macro-ecological pattern: i.e. the emergence of consistent patterns of variation of species richness and body size with ecosystem surface area, despite the heterogeneity and variability of their taxonomic composition.

The species–area patterns observed for the macro-invertebrate guilds of the Italian lagoons did not seem to be determined by bias related to the data mining activity and to the resulting data pool. The species–area patterns did not depend on the fact that larger lagoons could have attracted more research activities within an area, since there was no significant relationship between lagoon surface area and the number of papers retrieved from the literature (see data in Table 2). The fact that the same patterns were observed for the entire macro-invertebrate guild, for the dominant phyla and at different degrees of taxonomic resolution suggested that the observed patterns did not depend on any taxonomic bias in the papers selected, which could have been less selective for smaller lagoons, where it is easier to have more homogeneous sampling of all the environmental heterogeneity, nor did it depend on different taxonomic resolution among lagoons. Moreover, the observed patterns did not seem to be dependent on the criteria utilised for selecting papers among those retrieved with the data mining techniques, since the same patterns were observed in the two samples of 26 and 16 lagoons, which were selected according to slightly different criteria, more selective for the 16 lagoon sample.

The "species–area relationship" is a common pattern in insular ecosystems, both geographic islands (MacArthur and Wilson, 1967; Holt et al., 1999) and habitat islands (MacArthur, 1972). Transitional waters represent a particular type of habitat island, whose potential species pool is composed of the pool of freshwater and marine species tolerant to sharp variation gradients in the abiotic environment. Consequently, macro-invertebrate guilds of transitional waters do not represent proper meta-communities independent of the neighbouring freshwater and marine patches.

Different mechanistic explanations have been proposed to account for the species–area relationships, including a "sample effect" on the regional species pool (Connor & McCoy, 1979), and a scaling of immigration and extinction rates within a area based on habitat productivity (Wright, 1983), habitat heterogeneity (Williams, 1964) and neutral selection (He, 2005). Nevertheless, relationships between taxonomic richness and energy availability are among the best documented macro-ecological patterns (Wright, 1983; Hawkins et al., 2003; Evans et al., 2005). The More Individual Hypothesis (MIH, Wright, 1983), which stipulates positive species–energy relationships, has often been proposed as a mechanistic basis of the observed patterns. For the same reason, energy availability has also been proposed as a major mechanistic basis of the species–area relationships (Bonn et al., 2004). Coastal lagoons are generally very productive habitats per unit of surface and their overall productivity is very likely to scale with the lagoon surface area, at least when comparisons cover a wide spectrum of surface areas, such as in the present study (i.e. from 0.19 km^2 to 549 km^2). Therefore, the MIH would also apply to the lagoon ecosystems, even though a sampling effect could also have contributed. However, the consistency of the slopes of the species–area relationships within taxonomic phyla, which varied between 0.21 and 0.24, when compared with the very different taxonomic richness among phyla, which ranged from 94 taxa (Bivalvia) to 242 taxa (Others), seemed too strong to be explained by a stochastic sampling effect, while it would be perfectly consistent with a deterministic integration of the capacity rule (*sensu* Brown, 1981) set by energy availability and the allocation rule arising from species interactions and niche partitioning.

Larger species require larger home ranges (e.g. Basset, 1995), have higher metabolic rates per capita (Peters, 1983) and are more exposed to extinction risks, unless a sufficiently high population density is permitted by sufficient energy availability. Direct relationships of maximum body size of macro-invertebrate taxa occurring in benthic guilds and energy availability have already been observed (Basset, 1994). The body size–area relationships could also be explained by a sampling effect in a given species pool in which taxonomic richness increased with the lagoon surface area. On the other hand, even considering the body size–area relationships, the narrow range of slope variation, from 0.19 to 0.28, when compared with the ranges of the number of taxa for which standard body size was available in the literature, from 27 (Bivalvia) to 102 (Gastropoda), and of body size variability within phyletic groups, measured as the ratio of the largest to the smallest taxa, from 70 (Bivalvia) to 315 (Gastropoda), seemed more likely to be due to energetic deterministic factors and body size related coexistence conditions (Basset, 1995) than to a stochastic sampling effect. The observed maximum body size area relationships did not seem to be affected by some bias in the data set, since they were observed, albeit with different statistical significance, after grouping data in different ways and comparing different phyletic groups. The fact that these patterns were observed when using body length data obtained from the literature support their robustness, even though more direct experimental tests are required before generalising the observed patterns to macro-ecological patterns relating the surface area of transitional aquatic ecosystems to energy availability and body size of the largest species in the macro-invertebrate benthic guilds.

The macro-ecological patterns observed for macro-invertebrate guilds of Italian lagoons emphasised the insularity of these ecosystem patches in the coastal landscapes, despite their peculiarity of being patches surrounded by a space which is a suitable habitat rather than an unsuitable habitat, as the definition of island generally specifies (e.g. MacArthur, 1972), for all the lagoon macro-invertebrate taxa. Moreover, they suggested that species–area relationships in coastal lagoons were not particularly affected by immigration.

The observed macro-ecological patterns, i.e. taxonomic heterogeneity of macro-invertebrate guild compositions in Italian lagoons and the consistency of species–area and body size–area distributions of benthic macro-invertebrates, have some important implications for the monitoring of the transitional water ecological status, which represent a challenge for scientists (Basset & Abbiati, 2004), and for the implementation of WFD.

The species–area relationships underline the importance of physiographic characteristics (i.e. shape) of transitional water ecosystems in the identification of transitional water types, as required by the WFD. In fact, since species richness is one of the suggested descriptors for different biological quality elements in the WFD it seems to be necessary to take into account the scaling of taxonomic richness with lagoon surface area in order to minimise the intra-type variability of taxonomic richness, which is a major reason for splitting transitional water ecosystems into ecosystem types. Moreover, species–area relationships, i.e. the scaling of average taxonomic richness with lagoon surface, could also represent a standardisation tool for classifying the ecological status of transitional ecosystems, distinguishing good status, i.e. observed taxonomic richness higher than standard taxonomic richness, from poor status, i.e. observed taxonomic richness lower than standard. Nevertheless, this approach could make it very easy to distinguish, at the desired statistical level of significance, "moderate" status (in the terms of the WFD), i.e. standard richness with a proper confidence interval, from "good" or "poor" ecological status.

Finally, the observed patterns of body size area relationships would also have an important application to the field of monitoring transitional ecosystem health. In fact, body size is a relevant individual feature, which is relatively simple to measure, suitable for inter-calibration and comparable also among very different guilds, with an appropriate transformation and translation of the x-axis. Therefore, body size related descriptors are particularly suitable as potential descriptors of ecosystem health: i.e. they are consistently much less variable than taxonomic composition, yet they show clear patterns of variation with ecologically relevant parameters such as ecosystem surface area. These findings could open up new perspectives in the field of bio-monitoring, with important implications for the analysis of aquatic ecosystem health.

Acknowledgements This work was funded by a FIRB grant and an INTERREG IIIB CADSES grant to Alberto Basset and by an APAT grant to Letizia Sabetta. The authors thank Berenice Varvaglione for the support in the data mining process and Giorgio Mancinelli for his critical comments on a draft version of this paper.

References

Amarasinghe, U. S. & R. L. Welcomme, 2002. An analysis of fish species richness in natural lakes. Environmental Biology of Fishes 65: 327–339.

Andrefouet, S., M. Claereboudt, P. Matsakis, J. Pages & P. Dufour, 2001. Typology of atoll rims in Tuamotu Archipelago (French Polynesia) at landscape scale using SPOT HRV images. International Journal of Remote Sensing 22: 987–1004.

Arrenhius, O., 1921. Species and area. Journal of Ecology 9: 95–99.

Basset, A., 1994. Mole corporea ed organizzazione delle guild a base detrito all'interfaccia terra-acqua dolce. Atti Congresso S.IT.E. 16: 155–160.

Basset, A., 1995. Body size-related coexistence: an approach through allometric constraints on home-range use. Ecology 76: 1027–1035.

Basset, A. & M. Abbiati, 2004. Challenges to transitional water monitoring: ecological descriptors and scales. Aquatic Conservation: Marine and Freshwater Ecosystems 14: S1–S3.

Basset, A. & L. Sabetta, 2005. Identificazione e caratterizzazione degli ambienti acquatici di transizione: risultati di una indagine italiana a livello nazionale. Atti Workshop "Il monitoraggio delle acque di transizione". Venezia 27–29 ottobre 2004: 78–86.

Basset, A., L. Sabetta, A. Fonnesu, D. Mouillot, T. Do Chi, P. Viaroli, G. Giordani, S. Reizopoulou, M. Abbiati & G. C. Carrada, 2006. Typology in Mediterranean transitional waters: new challenges and perspectives. Aquatic Conservation 16: 441–455.

Battaglia, B., 1959. Final resolution of the symposium on the classification of brackish waters. Archivio Di Oceanografia E Limnologia 11(suppl): 243–248.

Bonn, A., D. Storch & K. J. Gaston, 2004. Structure of the species–energy relationship. Proceeding: Biological Sciences 271: 1685–1691.

Brown, J. H., 1981. Two decades of homage to Santa Rosalia: toward a general theory of diversity. American Zoologist 21: 877–888.

Carrada, G. C., F. Cicogna & E. Fresi, 1988. Le lagune costiere: ricerca e gestione. CLEM Pubbl, Napoli, 254 pp.

Comin, F. A., M. Menendez & J. A. Herrera, 2004. Spatial and temporal scales for monitoring coastal aquatic ecosystems. Aquatic Conservation: Marine and Freshwater Ecosystems 14: S5–S17.

Connor, E. F. & E. D. McCoy, 1979. The statistics and biology of the species–area relationship. The American Naturalist 113: 791–833.
Elmberg, J., P. Nummi, H. Pöysä & K. Sjöberg, 1994. Relationships between species number, lake size and resource diversity in assemblages of breeding waterfowl. Journal of Biogeography 21: 75–84.
Evans, K. L., J. J. D. Greenwood & K. J. Gaston, 2005. The role of extinction and colonisation in generating species–energy relationships. Journal of Animal Ecology 74: 498–507.
Gaston, K. J., 2000. Global patterns in biodiversity. Nature 405: 220–227.
Griffiths, D., 1997. Local and regional species richness in North American lacustrine fish. The Journal of Animal Ecology 66: 49–56.
Guelorget, O. & J. P. Perthuisot, 1983. Le domaine paralique. Expressions géologiques, biologiques et économiques du confinement. Travaux du laboratoire de géologie 16: 1–136.
Guelorget, O., G. F. Frisoni & J. P. Perthuisot, 1983. Zonation biologique des milieux lagunaires: definition d́une echelle de confinement dans le domaine paralique méditerranéen. Journal de Recherche oceanographique 8: 15–35.
Hawkins, B. A., R. Field, H. V. Cornell, D. J. Currie, J. F. Guegan, D. M. Kaufman, J. T. Kerr, G. G. Mittelbach, T. Oberdorff, E. M. O'Brien, E. M. Porter & J. R. G. Turner, 2003. Energy, water and broad scale patterns of species richness. Ecology 84: 3105–3117.
He, F., 2005. Deriving a neutral model of species abundance from fundamental mechanisms of population dynamics. Functional Ecology 19: 187–193.
Holt, R. D., J. H. Lawton, G. A. Polis & N. D. Martinez, 1999. Trophic rank and the species–area relationship. Ecology 80: 1495–1504.
Hubbell, S. P., 2001. The unified neutral theory of biodiversity and biogeography. Princeton University Press, Princeton, 448 pp.
Hurtt, G. C. & S. W. Pacala, 1995. The consequences of recruitment limitation. Reconciling chance, history and competitive differences between plants. Journal of Theoretical Biology 176: 1–12.
Istituto Idrografico della Marina, 1999. Tavola italiana delle maree, Genova, 118 pp.
Kjerfve, B., 1994. Coastal Lagoon Processes. Elsevier, Amsterdan, 577 pp.
Leibold, M. A., M. Holyoak, N. Moquet, P. Amarasekare, J. M. Chase, M. F. Hoopes, R. D. Holt, J. B. Shurin, R. Law, D. Tilman, M. Loreau & A. Gonzales, 2004. The metacommunity concept: a framework for multi-scale community ecology. Ecology Letters 7: 601–613.
MacArthur, R. H., 1970. Species packing and competitive equilibrium for many species. Theoretical Population Biology 1: 1–11.
MacArthur, R. H., 1972. Geographical Ecology. Harper & Row, N.Y., 269 pp.
MacArthur, R. H. & E. O. Wilson, 1967. The Theory of Island Biogeography. Princeton Univ. Press, Princeton, N.J., 203 pp.
Matter, S. F., I. Hanski & M. Gyllenberg, 2002. A test of the metapopulation model of the species–area relationship. Journal of Biogeography 29: 977–983.
Peters, R. H., 1983. The Ecological Implications of Body Size. Cambridge University Press, Cambridge, 344 pp.
Rosenberg, D. M. & V. H. Resch, 1993. Freshwater Biomonitoring and Benthic Macroinvertebrates. Chapman & Hall, New York, N.Y., 488 pp.
Sabetta, L., E. Barbone, A. Giardino, B. Varvaglione, C. Silvestri, N. Galuppo, G. Casazza & A. Basset, 2005. An approach to typology and classification of ecological status of transitional waters. UNESCO Proceeding Book: 76–83.
Schoener, T. W. & D. A. Spiller, 1992. Is extinction rate related to temporal variability in population size? An empirical answer for orb spiders. American Naturalist 139: 1176–1207.
Simberloff, D. S. & L. G. Abele, 1976. Island biogeography theory and conservation practice. Science 191: 285–286.
Srivastava, D. S. & J. H. Lawton, 1998. Why more productive sites have more species: an experimental test of theory using tree-hole communities. American Naturalist 152: 510–529.
Williams, C. B., 1964. Patterns in the Balance of Nature. Academic Press, London, 324 pp.
Wright, D. H., 1983. Species–energy theory: an extension of species area theory. Oikos 41: 496–506.

Hydrobiologia (2007) 577:141–149
DOI 10.1007/s10750-006-0423-6

Index of size distribution (ISD): a method of quality assessment for coastal lagoons

Sofia Reizopoulou · Artemis Nicolaidou

Abstract A new index was developed as a tool for quantifying the degree of disturbance in lagoons in order to meet the objective of Ecological Quality Status (EcoQ), using the zoobenthos quality element. The Index of Size distribution (ISD) is proposed to assess the ecological quality status of coastal lagoons. It represents the skewness of the distribution of individuals of a benthic community in geometric size (biomass) classes. The ISD was applied in three coastal lagoons with different levels of disturbance and classified them as of good, moderate and poor ecological quality. A scheme for the classification of EcoQ in lagoonal systems is presented. The index showed a strong relationship with the percentage of organic carbon in the sediment, as well as with the dissolved oxygen concentrations. ISD having the advantage of good discriminating power and not demanding high taxonomic resolution, could be a simple and promising tool to be further applied and tested in Mediterranean lagoons.

Keywords Lagoons · Benthos · Pollution assessment · Indices · Ecological quality status

Introduction

Coastal lagoons are shallow, relatively enclosed water bodies. They can be considered as harsh, naturally stressed environments, characterised by frequent fluctuations of environmental parameters on a daily and seasonal basis. This natural instability discourages the settlement of many species, resulting in a low number of species and low diversity. On the other hand, they are organically enriched areas, both as a result of the riverine inputs and the recycling of materials within the system, thus a large number of individuals, summing high biomass values, is attained.

The above characteristics of the lagoons would be rather indicative of a polluted situation in the marine environment, especially in the oligotrophic Eastern Mediterranean, which is generally characterised by low abundance and high diversity (Bellan-Santini, 1985). Therefore methods used to assess pollution in the marine environment may not be applicable to the lagoons (Reizopoulou et al., 1996).

Guest editors: P. Viaroli, P. Lasserre & P. Campostrini
Lagoons and Coastal Wetlands in the Global Change Context: Impacts and Management Issues

S. Reizopoulou (✉)
Institute of Oceanography, Hellenic Centre for Marine Research, PO Box 712, 190 13 Anavissos, Attiki, Greece
e-mail: sreiz@ath.hcmr.gr

A. Nicolaidou
Department of Zoology – Marine Biology, School of Biology, University of Athens, 157 84 Panepistimiopoli, Athens, Greece

European Union Water Framework Directive (2000/60/EC) requires that the member states establish ecological quality classification systems for all surface waters, including transitional waters. The recently developed biotic indices of ecological quality AMBI (Borja et al., 2000; 2003) and BENTIX (Simboura & Zenetos, 2002), are based on the concept of indicator species and are suitable for assessing EcoQ in coastal waters. However, since these indices use ecological groups of species according to their sensitivity to stress, they should be used with caution in lagoons, which are ecosystems naturally inhabited by species able to tolerate stressed conditions.

The application of body-size distribution is an alternative method to investigate benthic community structure. Changes of benthic community biomass under disturbed conditions are well documented in benthic ecology (Pearson & Rosenberg, 1978; Warwick, 1986). The increasing organic pollution results in loss of the larger long-lived species (k-strategists) from the community in favour of more tolerant short-lived opportunists (r-strategists) (Pearson & Rosenberg, 1978). The former dominate in terms of biomass, the latter in terms of abundance.

In the present study, an index (Index of Size Distribution – ISD) was applied to the macroinvertebrates of three Greek coastal lagoons with different degrees of pollution. ISD is an alternative taxonomic free method, developed for lagoons, based on the distribution of individuals of benthic communities in biomass size classes. The skewness of the distribution was used as a measure of disturbance and a classification scheme of environmental quality is proposed.

Materials and methods

Sampling sites

Samplings were performed in three Greek brackish water lagoons, with different degrees of disturbance (Fig. 1). Tsopeli lagoon, is situated at the mouth of River Louros in Amvrakikos Gulf (Ionian Sea) and has no obvious source of pollution. On the other end of the pollution scale is Papas lagoon in SW Peloponnisos, communicating with both the Patraikos Gulf and the Ionian Sea. It is an organically polluted ecosystem where anoxic crises are known to occur. The organic carbon, sulphur and phosphorus concentrations were found significantly elevated in the surface sediments of the Papas lagoon. Also heavy metals presented high values compared to other lagoons (Kaberi et al., 2000).

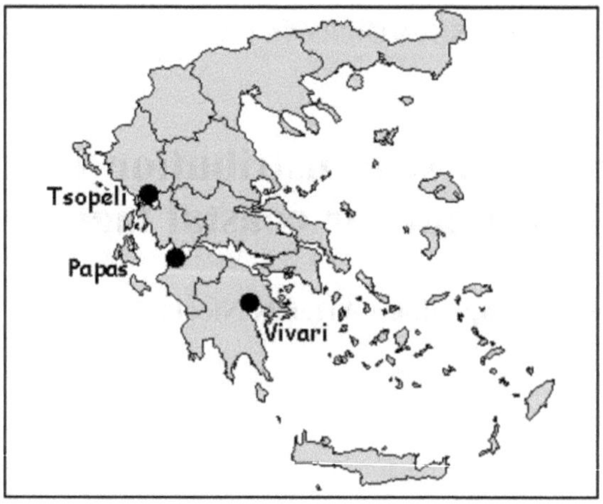

Fig. 1 Sampling sites

In the middle of the scale, Vivari lagoon, in Argolikos Gulf, receives a small intermittent flow of fresh water from a runnel and no obvious source of pollution was observed at the time of sampling. However, a sudden disappearance of vegetation was reported 1 year earlier.

All the lagoons are shallow systems with depths around 0.5 m, reaching 1.5 m only locally. Narrow barriers isolate them from the sea, communicating through small openings. They are used for extensive (Tsopeli and Vivari) and semi-intensive (Papas) fish farming.

Tsopeli lagoon is characterized by the presence of angiosperms (*Zostera noltii*), while in Papas lagoon enormous amounts of decomposing *Ulva rigida* are responsible for dystrophic crises. The prolonged anoxic events in the southern part of the lagoon, often lead to release of hydrogen sulphide, with consequent massive mortality of fish and clams.

Sampling and laboratory methods

A dense grid of stations was sampled for abiotic parameters in order to acquire a detailed picture of the physicochemical conditions in the studied lagoons. Salinity, temperature and dissolved oxygen were measured in situ, using Yellowspring probes.

Macrozoobenthos was sampled seasonally: five times in Tsopeli and Vivari lagoons in 1990–1991, and five times in Papas lagoon in 1998–1999. Six stations were visited in Tsopeli, four in Vivari and three in Papas lagoon. The biomass data of benthic communities were available only for four stations in Tsopeli and three in Vivari.

Macrofaunal samples were collected using a Ponnar grab (0.05 m^2) and three replicate samples were collected from each site. The samples were sieved through a 1-mm mesh, stained with Rose Bengal and preserved in 4% formalin. In the laboratory, the macrofauna was sorted, identified at species level and counted.

A sub-sample of sediment was used for granulometry and organic carbon analysis. Organic carbon analysis was carried out according to Gaudette et al. (1974) for Tsopeli and Vivari and according to Verardo et al. (1990) in Papas lagoon.

For the determination of the ISD the individual body size was expressed as body weight (mg). Individual body weight of the animals was obtained after drying at 60°C for 48 h and weighing at the 0.0001 g level. The polychaetes were removed from their tubes and mollusc shells were dissolved with dilute hydrochloric acid prior to biomass determination.

To examine the distribution of individuals per geometric size classes (class I = 0.1 mg, class II = 0.2–0.3 mg, class III = 0.4–0.7 mg, ... class XII = 204.8–409.5 mg), histograms were plotted presenting the percentage of individuals belonging to each geometric size class for each station. For every size-distribution set, a skewness value was calculated and the ISD classification scheme was produced, by plotting the whole series of skewness values obtained.

Multivariate and univariate analyses were performed using the program PRIMER-E 2000.

Results

Environmental variables

The ranges of abiotic variables for each lagoon are shown in Table 1. Salinity and temperature showed a wide range of values as a result of the lagoon shallowness and the degree of confinement (Table 1). Smaller ranges were observed in Vivari, which had the highest degree of communication with the sea.

In Tsopeli oxygen concentrations were enhanced by the presence of phanerogam meadows, whilst the sedimentary organic carbon presented the lowest values (Table 1). The Vivari lagoon had a bare sediment with a high organic carbon content. In the Papas lagoon the sedimentary organic carbon was also high. Moreover, in the southern part persistent summer anoxic events occurred, due to the mass development and further decomposition of *U. rigida* biomasses.

Community attributes

The differences in environmental characteristics of the lagoons were reflected in their benthic communities. The MDS of Fig. 2 grouped the stations of each lagoon according to their faunal similarities.

Tsopeli was characterised by species typical of brackish water lagoons, the most dominant of which were *Abra ovata, Cerastoderma glaucum, Mytilaster minimus* and larvae of Chironomidae insects. The abundance of the polychaetes *Nephtys hombergi* and *Heteromastus filiformis* was also high, while the crustaceans *Gammarus insensibilis* and *Idotea baltica* were found within the vegetation stands.

Table 1 Range of abiotic variables in each lagoon

Lagoon	Depth (m)	*S* (psu)	*T* (°C)	O_2 (mg l^{-1})	Coarse material (%)	Organic C (%)
Tsopeli	0.2–1.5	21.0–38.0	8.0–29.0	2.8–9.8	6.7–66.3	1.1–5.3
Vivari	0.6–1.5	28.5–40.0	12.0–34.0	4.2–8.4	16.4–35.1	3.1–6.7
Papas	0.2–1.5	20.0–42.5	10.0–32.0	0.8–9.3	23.0–98.0	2.9–5.6

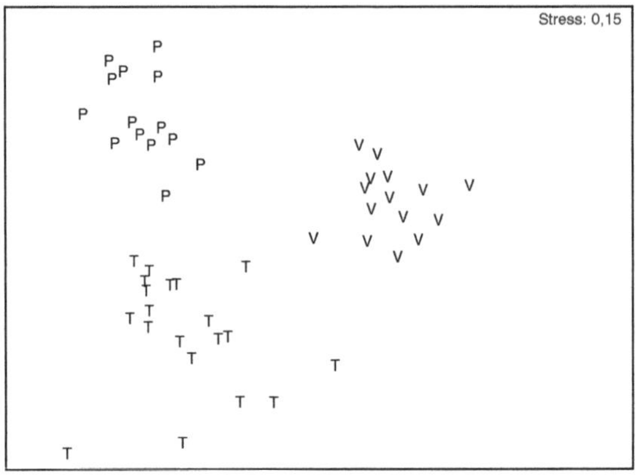

Fig. 2 Multidimensional scaling based on the species abundances [log (*x* + 1)]

In Vivari two species, *Abra ovata* and *Heteromastus filiformis* alternated in dominance. In autumn *Abra ovata* decreased in favour of *Hediste diversicolor*. Other species of molluscs and crustaceans were almost absent. Finally, in Papas lagoon the bivalve *Abra ovata* and the serpulid *Hydroides dianthus* were very abundant. High densities of amphipods (*Corophium insidiosum, Microdeutopus gryllotalpa*) and opportunistic polychaetes (*Capitella capitata, Heteromastus filiformis*) were observed seasonally. In summer, the southern part of the lagoon became azoic due to the anoxia, while in the northern part clam populations (*Tapes decussatus*) disappeared.

The variations of macrobenthic community characteristics, namely number of species (*S*), diversity (*H′*), evenness (*J*) and abundance (*N*) of each lagoon are shown in Table 2. The highest number of species was found in Tsopeli (84) and the lowest in Vivari (64). The diversity neither varied according to the degree of disturbance in the lagoons, nor showed a statistically significant correlation with dissolved oxygen concentrations in the water column and organic carbon in the sediment. The lowest diversity value was found in Vivari, characterised by total absence of vegetation, while the highest number of individuals was noted in Papas, the most eutrophicated lagoon.

The index of size distribution

The frequency distribution of geometric size (biomass) classes was plotted for all stations and seasons in the three lagoons. Examples of the distributions are shown in Fig. 3. It is evident that the undisturbed conditions correspond to more even distribution of smaller and larger size classes, while under disturbed conditions an uneven distribution of the size classes is obvious, with the smaller ones being the most abundant. This can be expressed numerically by the skewness of distribution (the novel feature of this index).

The differences in size distribution were not only due to the presence of small opportunistic species in Vivari and Papas. Some of the most abundant species attained a larger size in the less disturbed Tsopeli, as indicated by the mean individual size of *Abra ovata*, *Cerastoderma glaucum*, *Tapes decussatus* and *Hediste diversicolor* in Fig. 4.

The plot of Fig. 5 illustrates the range of the skewness values over the lagoons studied. As with the environmental parameters and the fauna, the ISD values varied in each lagoon. In Tsopeli the index ranged from high (–0.34) to moderate (2.51), in Vivari from good (1.85) to poor (3.28) and in Papas the ISD was even higher ranging from moderate (2.26) to poor (3.45). Figure 6 shows the mean ISD for each lagoon. Overall, Tsopeli showed the lowest mean ISD value, classifying it as belonging to 'good' ecological class, while ecological quality in Vivari and Papas was characterised as 'moderate' and 'poor', respectively.

Table 2 Total number of species and variations of community features in each lagoon

Lagoon	Total no. of species *S*	Variations of no. of species *S*	Diversity *H′*	Evenness *J*	Abundance *N*
Tsopeli	84	5–45	1.3–3.7	0.48–0.89	508–5827
Vivari	64	7–21	0.7–3.4	0.21–0.78	753–8820
Papas	76	0–44	1.7–3.7	0.45–0.79	0–44108

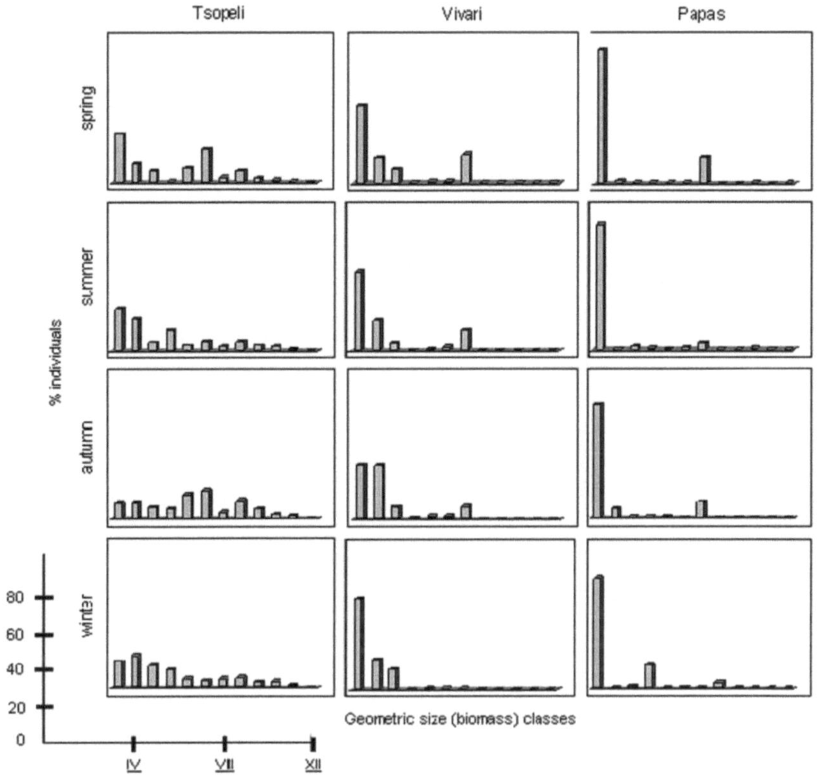

Fig. 3 Examples of size distributions of benthic communities in the studied lagoons

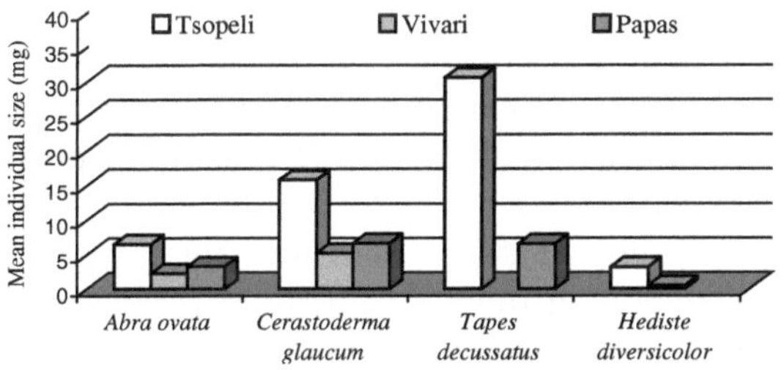

Fig. 4 Mean individual size of some abundant species in the three lagoons

A scheme for the classification of Ecological Quality Status in lagoonal systems is presented in Table 3.The boundary limits among classes were set following a linear scale and according to the plot. The respective Ecological Quality Ratio (EQR), defined as the ratio of the observed value versus the value of the metric under reference conditions (EC, 2003) is also given in Table 3. The EQR values are standardized to fit the 0–1 range.

Validation of the method

Figure 7 shows the regression between the ISD and the percentage of the organic carbon in the sediment. The significant correlation ($r = 0.63$, $p = 0.0000$) indicates that the ISD co-varies with the organic pollution gradient in the lagoons. The ISD was significantly correlated with dissolved oxygen concentrations ($r = -0.35$, $p = 0.0001$), which is also a measure of environmental health.

Figure 8 shows the mean ISD values within each ecological class against the corresponding mean organic values. There is a strong correspondence between ISD and organic carbon values along the ecological classes defined by the metric.

Discussion

The most common and serious anthropogenic impact in Mediterranean coastal lagoons is nutri-

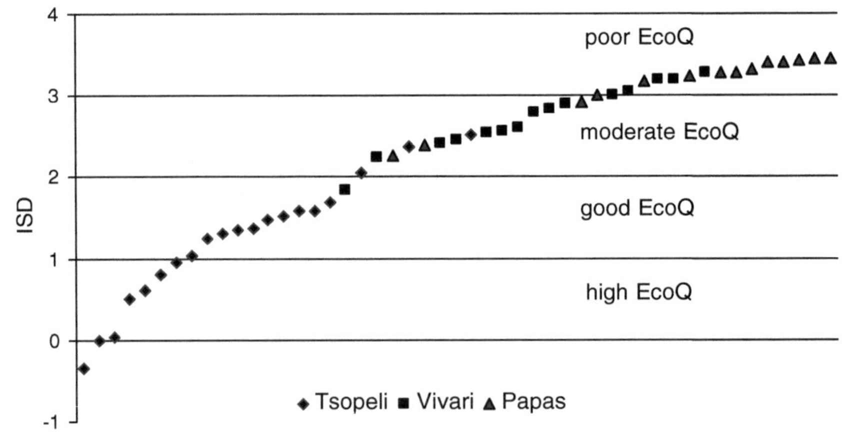

Fig. 5 Values of ISD, as calculated for the three lagoons and the corresponding proposed EcoQ classification

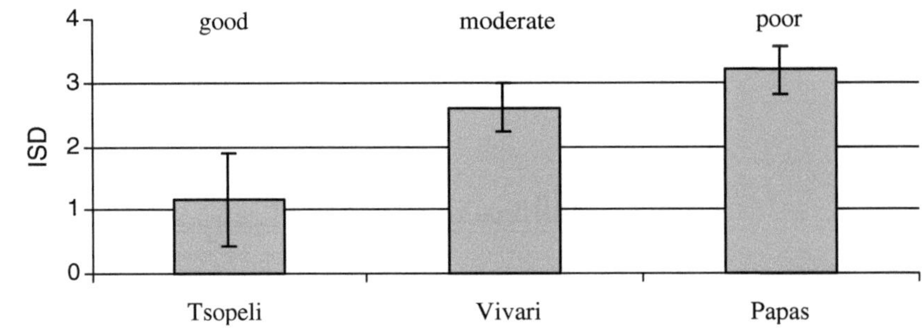

Fig. 6 Mean values and standard deviation of ISD in each lagoon

Table 3 Classification scale of Ecological Quality Classes based on ISD

EcoQ	ISD	EQR
High	$-1 \le$ ISD < 1	1
Good	$1 \le$ ISD < 2	0.60
Moderate	$2 \le$ ISD < 3	0.39
Poor	$3 \le$ ISD < 4	0.20
Bad	Azoic conditions	0

ent enrichment, which often leads to a replacement of sea-grasses by opportunistic green macroalgae (Valiela et al., 1997; Orfanidis et al., 2001) and leads to oxygen depletion known as 'dystrophic crises' (Sfriso et al., 1992; Viaroli et al., 1996). Temporal variations in benthic communities associated with such eutrophication phenomena have been the subject of numerous studies (e.g., Lardicci et al., 1997, 2001; Tagliapietra et al., 1998; Koutsoubas et al., 2000). According to most of the above authors, increased organic disturbance results in an increase of opportunistic and tolerant lagoonal species, in an increase of densities, and in a decline of suspension feeders and carnivores in favour of sub-surface deposit feeders. Pearson & Rosenberg (1978) suggested that the average individual size decreases in polluted areas. Under disturbed conditions the larger, long-lived species are the first to disappear and the communities are dominated, by smaller, short-lived opportunistic species.

The results of the present investigation are in accordance with the abovementioned comments. Biomass profile may highlight alterations on benthic ecosystem along a pollution gradient, through intense changes on community size structure. Small-bodied invertebrates may characterize environments with high instability and small body size could be a consequence of the environmental/anthropogenic pressures imposed on the organisms. The small size classes of the studied communities, were dominated by tolerant and opportunistic deposit feeders, while the larger size classes were mostly dominated by filter feeding bivalves and carnivorous polychaetes. Algal blooms, anoxic events and sulphide production in Papas lagoon depleted the abundant filter feeders, lowered the abundance of

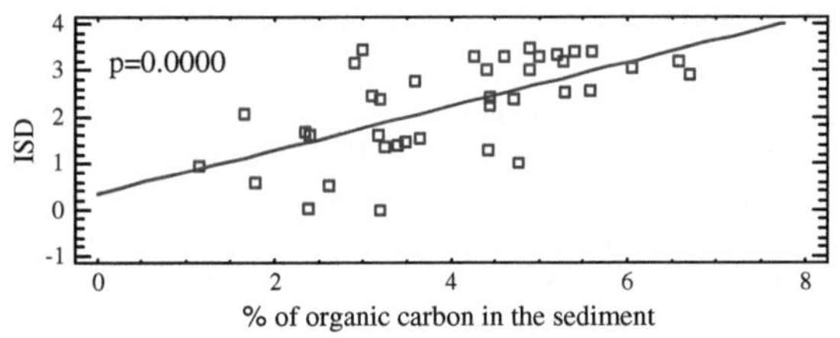

Fig. 7 ISD against percentage of organic carbon in the sediment

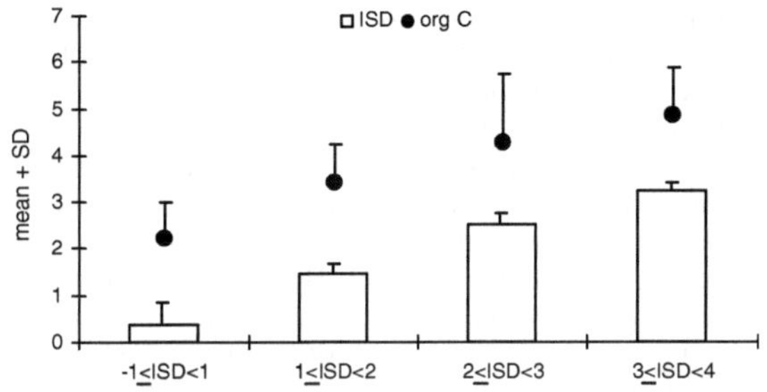

Fig. 8 Mean and standard deviation of ISD and organic carbon values within each ecological class

many sedentary species, and, at the same time, allowed a stronger representation of mobile small bodied grazers (crustaceans) able to withstand the constantly disturbed environment. The dominance of tolerant species in lagoons mainly indicates the natural instability of the environment, while the disappearance of populations of large filter feeding bivalves such as *Tapes decussatus* in Papas Lagoon, should raise concern for the community health.

A large number of methods proposed to assess degradation of the marine environment are based on benthic communities. Originally, these methods were developed using data collected from marine areas, and their applicability in coastal Mediterranean lagoons was first questioned by Reizopoulou et al. (1996). The natural environmental stress precludes the graphical method of Gray & Mirza (1979) based on the distribution of species in geometric abundance classes, since it relies on the fact that unstressed communities host many rare species while a small number of opportunists dominate; on the other hand in stressed environments the rare species are eliminated and many opportunists become extremely abundant. The ABC method (Warwick, 1986) and W statistic (Clarke, 1990) were successful in discriminating among impacted and not impacted lagoons in some occasions (Reizopoulou et al., 1996) but in some others were not (Lardicci & Rossi, 1998).

Regarding the community diversity, used widely as an index of environmental quality (Rosenberg et al., 2004), it should be noted that it cannot be successfully used in lagoons. Here, the natural instability and organic enrichment create extreme conditions where few species can be established and where diversity, depending on species richness and evenness of distribution, remains naturally low. Indeed, Reizopoulou & Nicolaidou (2004) found a strong negative correlation between diversity and confinement (sensu Guelorget & Perthuisot, 1983, 1992), as instability of environmental conditions increases with increasing isolation from the sea. Nevertheless confinement is a natural situation not always associated with environmental health. Arvanitidis et al. (2005) tested the rapid biodiversity assessment techniques on a pan-Mediterranean scale and found that although these techniques can reveal biodiversity patterns they are, nevertheless, inadequate for distinguishing naturally disturbed lagoons from anthropogenically impacted at a regional scale.

Reizopoulou et al. (1996) suggested that methods which use biomass are more reliable than those based on abundance. According to the literature, biomass structure is an important attribute of the community. Edgar (1994) found that size (biomass) structure of macrofaunal communities varied consistently between assemblages associated with macroalgal habitats of different morphology. This author suggested that the existence of relationships between community body size and environmental parameters might provide insight into the functioning of benthic communities. Jennings et al. (2002) demonstrated that there is a significant relationship between body weight and trophic level and suggested that analysis of temporal and spatial changes in size spectra could be used to detect temporal and spatial changes in trophic structure and to assess the impact of disturbance. Finally, Basset et al. (2004) who discussed the advantages and disadvantages of benthic macroinvertebrate body size descriptors as a tool for environmental monitoring, suggest that body size abundance distribution is related to disturbance pressure through individual energetics, population dynamics, interspecific interactions and species coexistence responses.

In the present study, the ISD based on biomass showed good correlation with the organic carbon in the sediment and the dissolved oxygen, two parameters related to environmental degradation. The EcoQ gradient illustrated by the ISD index is syntonic with the organic carbon gradient (Fig. 8).

ISD index seems to be a promising approach and a simple and effective tool for the ecological quality assessment of coastal lagoons. The new index has to be applied in other transitional water ecosystems, in order to set the confidence intervals of the boundary limits across the five EcoQ levels. It is important to focus on some points when applying the index. Given the high spatial variability of physical and chemical factors in lagoons, the ecological status may vary significantly. Furthermore, an intense seasonal variation is expected due to reproduction patterns: recruitment, for example, would tend to increase the skewness of the biomass distribution. Thus, in order to define the integrated ecological status for a given lagoon, the mean value of the index at various instances in space and time should be used.

The development of indicators and metrics is highly driven by the obligation of the European countries to meet the WFD requirements to classify the ecological status in coastal and transitional waters. Tools that are simple, practical, robust and cost effective (Rapid Assessment Techniques – RATs) are highly valued under the perspective of establishing monitoring and management plans.

The greatest advantage of the ISD over other indices is that it does not require high taxonomic resolution of the fauna, which is an extremely costly and time consuming process. The animals are weighed individually, independently of the species to which they belong. This makes the ISD a very practical tool for monitoring and management of the harsh, but at the same time fragile lagoonal ecosystems.

References

Arvanitidis, C., G. Chatzigeorgiou, D. Koutsoubas, C. Dounas, A. Eleftheriou & P. Koulouri, 2005. Mediterranean lagoons revisited: weakness and efficiency of the rapid biodiversity assessment techniques in a severely fluctuating environment. Biodiversity and Conservation 14: 2347–2359.

Basset, A., F. Sangiorgio & M. Pinna, 2004. Monitoring with benthic macroinvertebrates: advantages and disadvantages of body size descriptors. Aquatic Conservation Marine and Freshwater Ecosystems 14: S43–S58.

Bellan-Santini, D., 1985. The Mediterranean benthos: reflections and problems raised by a classification of the benthic assemblages. In Moraitou-Apostolopoulou, M. & V. Kiortsis (eds), NATO Conference Series (Ser I Ecology) 8: 19–56.

Borja, A., J. Franco & V. Perez, 2000. A marine biotic index to establish the ecological quality of soft-bottom benthos within European estuarine and coastal environments. Marine Pollution Bulletin 40: 1100–1114.

Borja, A., I. Muxika & J. Franco, 2003. The application of a marine biotic index to different impact sources affecting soft-bottom benthic communities along European coasts. Marine Pollution Bulletin 46: 835–845.

Buchannan, J. B., 1984. Sediment analysis. In Holme, N. A. & A. D. McIntyre (eds), Methods for the study of benthos. Blackwell, Oxford, 41–65.

Clarke, K. R., 1990. Comparison of dominance curves. Journal of Experimental Marine Biology and Ecology 138: 143–157.

EC, 2000. Directive of the European parliament and of the Council 2000/60/EC establishing a framework for community action in the field of Water Policy. PE-CONS 3639/1/00.

EC, 2003. Guidance on typology, reference conditions and classification systems for transitional and coastal waters. Produced by CIS Working Group 2.4. (Coast), Common Implementation Strategy of the Water Framework Directive, European Commisssion, p. 116.

Edgar, G. J., 1994. Observations on the size-structure of macrofaunal assemblages. Journal of Experimental Marine Biology and Ecology 176: 227–243.

Gaudette, H. E., W. R. Flight, L. Toner & D. W. Folger, 1974. An inexpensive titration method for the determination of organic carbon in recent sediments. Journal of Sedimentary Petrology 44: 249–253.

Gray, J. S & F. B. Mirza, 1979. A possible method for detecting pollution induced disturbance on marine benthic communities. Marine Pollution Bulletin 10: 142–146.

Guelorget, O. & J. P. Perthuisot, 1983. Le domain paralique. Expression geologique, biologique du confinement. Travaux du Laboratoire de Geologie, Ecole Normale Superieure de Paris 16, Paris, 136 pp.

Guelorget, O. & J. P. Perthuisot, 1992. Paralic ecosystems. Biological organization and functioning. Vie et Milieu 42: 215–251.

Jennings, S., J. K. Pinnegar, N. V. C. Polunin & K. J. Warr, 2002. Linking size-based and trophic analyses of benthic community structure. Marine Ecology Progress Series 226: 77–85.

Kaberi, E., C. Anagnostou, E. Krassakopoulou, M. Taxiarhi & A. Papageorgiou, 2000. The quality of bottom sediments in Papas lagoon and their role to the ecosystem's supply with C, N, P and S. In Pagou, K. (ed.), Monitoring of the ecosystem Papas Lagoon Araxos/Achaia and suggestions for the management and protection of the area. NCMR, 2000 Technical Report, 149–164 (In Greek).

Koutsoubas, D., C. Dounas, C. S. Arvanitidis, S. Kornilios, G. Petihakis, G. Triantafyllou & A. Eleftheriou, 2000. Macrobenthic community structure and disturbance assessment in Gialova Lagoon, Ionian Sea. Journal of Marine Sciences 57: 1472–1480.

Lardicci, C., F. Rossi & A. Castelli, 1997. Analysis of macrozoobenthic community structure after severe dystrophic crises in a Mediterranean coastal lagoon. Marine Pollution Bulletin 34: 536–547.

Lardicci, C. & F. Rossi, 1998. Detection of stress on macrozoobenthos: evaluation of some methods in a coastal Mediterranean lagoon. Marine Environmental Research 45: 367–386.

Lardicci, C., S. Como, S. Corti & F. Rossi, 2001. Recovery of the macrozoobenthic community after severe dystrophic crises in a Mediterranean coastal lagoon (Orbetello, Italy). Marine Pollution Bulletin 42: 202–214.

Orfanidis, S., P. Panayotidis & N. Stamatis, 2001. Ecological evaluation of transitional and coastal waters: a marine benthic macrophytes-based model. Mediterranean Marine Science 2: 45–65.

Pearson, T. H & R. Rosenberg, 1978. Macrobenthic succession in relation to organic enrichment and pollution of the marine environment. Oceanography and Marine Biology Annual Review 5: 229–311.

Reizopoulou, S., M. Thessalou-Legaki & A. Nicolaidou, 1996. Assessment of disturbance in Mediterranean lagoons: an evaluation of methods. Marine Biology 125: 189–197.

Reizopoulou, S. & A. Nicolaidou, 2004. Benthic diversity of coastal brackish-water lagoons in western Greece. Aquatic conservation: Marine and Freshwater Ecosystems 14: S93–S102.

Rosenberg, R., M. Blomqvist, H. C. Nicolsson, H. Cederwall & A Dimming, 2004. Marine quality assessment by use of benthic species-abundance distributions: a proposed new protocol within the European Union Water Framework Directive. Marine Pollution Bulletin 49: 728–739.

Sfriso, A., B. Pavoni, A. Macromini & A. Orio, 1992. Macroalgae, nutrient cycles, and pollutants in the lagoon of Venice. Estuaries 15: 517–528.

Simboura, N. & A. Zenetos, 2002. Benthic indicators to use in Ecological Quality classification of Mediterranean soft bottom marine ecosystems, including a new Biotic Index. Mediterranean Marine Science 3: 77–111.

Tagliapietra, D., M. Pavan & C. Wagner, 1998. Macrobenthic community changes related to eutrophication in Pallude della Rosa (Venetian lagoon, Italy). Estuarine, Coastal and Shelf Science 47: 217–226.

Valiela, I., J. McClelland, J. Hauxwell, P. J. Behr, D. Hersh & K. Foreman, 1997. Macroalgal blooms in shallow estuaries: controls and ecophysiological and ecosystem consequences. Limnology and Oceanography 42: 1105–1118.

Verardo, D. J., P. N. Froelich & A. McIntyre, 1990. Determination of organic carbon and nitrogen in marine sediments using the Carlo Erba NA-1500 Analyzer. Deep-Sea Research 37: 157–165.

Viaroli, P., M. Naldi, C. Bondavalli & S. Bencivelli, 1996. Growth of the seaweed *Ulva rigida* C Agardh in relation to biomass densities, internal nutrient pools and external nutrient supply in the Sacca di Goro lagoon (Northern Italy). Hydrobiologia 329: 93–103.

Warwick, R. M., 1986. A new method for detecting pollution effects on marine macrobenthic communities. Marine Biology 92: 557–562.

Hydrobiologia (2007) 577:151–159
DOI 10.1007/s10750-006-0424-5

LAGOONS AND COASTAL WETLANDS

Low variation at allozyme loci and differences between age classes at microsatellites in grass goby (*Zosterisessor ophiocephalus*) populations

Paolo Maria Bisol · Alessandra Gallini · Sabrina Prevedello · Enza Rianna · Ezio Bernardinelli · Anita Franco · Lorenzo Zane

Abstract Allozymes and microsatellites were used to assess the level and distribution of genetic variation in grass goby population samples collected from the Venice Lagoon between October 2001 and May 2002. Eighteen enzymatic loci were examined in 434 individuals, 14 of which resulted to be monomorphic, and 4 (GPI-B*, LDH-B*, PGM-A*, PGM-B*) showed 2 alleles scored in 6 individuals only. Comparison with previous data suggests that genetic variation has been eliminated in the Venice Lagoon population during the last few years at three loci. In contrast, analysis of 11 microsatellites in a subset of 192 individuals revealed substantial molecular variation. Analysis of molecular variance showed a lack of genetic differentiation inside the lagoon with respect to site and date of collection, sex, and level of pollution. Significant variation in allelic frequencies was found at microsatellite loci when small (one year old) males were compared to large males (two and three years old), suggesting that a complex population dynamics occurs in this species. The very low level of polymorphism of allozymes could be due to the evolutionary history of the species, or, considering the difference between small and large males, could be the result of recent effects of drift. The second hypothesis is supported by the comparison with previous allozyme studies of the species in the same area, that suggests that loss of heterozygosity at three loci occurred in the last 10 years.

Keywords Genetic variability · Gobiidae · Adriatic · Allozymes · Microsatellites

Guest editors: P. Viaroli, P. Lasserre & P. Campostrini
Lagoons and Coastal Wetlands in the Global Change Context: Impacts and Management Issues

P. M. Bisol (✉) · A. Gallini · S. Prevedello · E. Rianna · E. Bernardinelli · L. Zane
Department of Biology, University of Padua, Via Giuseppe Colombo 3, 35131 Padova, Italy
e-mail: paolo.bisol@unipd.it

A. Franco
Department of Environmental Sciences, University of Venice, Celestia, Castello 2737/b, 30122 Venezia, Italy

Introduction

Natural selection and genetic drift can produce variation of allele frequencies in space and time. For this reason, genetic approaches may be interesting in the aquatic toxicology context, where they allow to investigate the effect of genetic diversity on biomarker response and to study the relationships between genetic polymorphism, pollution and demographic output.

In general, population genetics studies seem to suggest a higher than expected complexity of fish populations (Carvalho & Hauser, 1998). For example, despite the potential for dispersal, particularly at the larval stage, these organisms can often achieve a strong genetic differentiation at neutral loci, even on a local scale (Ward et al., 1994; Shaklee & Bentzen, 1998). Examples of genetic differentiation at the micro-geographic scale include differentiation between adjacent islands even in species with extended larval duration (Taylor & Hellberg, 2003), differentiation between adjacent estuaries (Ikeda et al., 2003; Maltagliati et al., 2003), and differentiation within the same lagoon (Planes et al., 1998). Moreover, investigation of temporal changes in allele frequencies highlighted the existence of genetic heterogeneity between cohorts in population samples taken from the same area (Ruzzante et al., 1996). This may be related to the complexity of recruitment in fish populations and to the large variance in reproductive success (Hedgecock et al., 1992), that can lead to over representation of individuals of the same family group. This temporal diversity may also result in long-term variation trends (Hutchinson et al., 2003). Both local genetic differentiation and temporal changes in the genetic constitution of populations can act as confounding factors in aquatic toxicology studies, because they can induce differences in biomarkers response caused by genetic differences between individuals rather than by exposure to pollutants.

The grass goby *Zosterisessor ophiocephalus* (Pallas, 1811) is a euryhaline and eurytherm species occupying inshore and brackish waters used as bioindicator. The species is widely distributed along the coast and in the lagoons and estuaries of the Mediterranean basin, Black Sea and Sea of Azov. In the Northern Adriatic lagoons, *Z. ophiocephalus* is an important component of the fish community, being widely distributed and present in all seasons (Mainardi et al., 2002). The species is preferentially found in areas where the sand bottom is covered by seagrass meadows and by macroalgal patches, and this preference is linked to the reproductive *habitus* (Marconato et al., 1996; Ota et al., 1996). Larger/older males burrow and defend a cavity nest in soft bottom sediments to attract and monopolise females. The reproductive strategy is made more complex by smaller/younger, opportunistic males (sneakers), anatomically and physiologically different from the territorial males, which penetrate into the subterranean nests and fertilize the laid eggs (Scaggiante et al., 1999; Mazzoldi et al., 2000; Torricelli et al., 2000).

Only few studies have been published on the genetics of *Z. ophiocephalus*. Callegarini & Ricci (1973) have analysed variability of the LDH locus, while more recent studies focused on phylogeny of Mediterranean gobies using both biochemical (McKay & Miller, 1991; Miller et al., 1994; Sorice & Caputo, 1999) and mitochondrial markers (Penzo et al., 1998).

This paper reports the results obtained for 18 gene–enzyme systems and 11 microsatellites, in 17 population samples of *Z. ophiocephalus*. Samples were collected in two consecutive years from sites of the Venice Lagoon showing different levels of pollution. Our aim is to provide aquatic toxicologists with baseline information about genetic variation in this species. In particular, the specific goals are: (1) to measure the level of genetic variation in grass goby in the Venice Lagoon, (2) to compare variation in the two different types of genetic markers, and, (3) to relate the pattern of genetic variation with site and date of collection, level of pollution, and with sex and length of analysed specimens.

Materials and methods

Samples collection

A total of 434 grass goby, representing 17 different samples, were collected in the Venice Lagoon during 2001 and 2002 (Table 1, Fig. 1). The sampling sites were classified by environmental risk classes (ERC) reported in Critto & Marcomini (2001). ERC is represented by the number of contaminants (from 1 to 9 in this case) above the benchmark Threshold Effect Level (TEL); at concentrations below TEL contaminants are expected to have no toxic effects on aquatic life.

Fish sampling techniques and general ecological characteristics are described in Franco et al. (2002). The station Ca′ Zane is affected by the

Table 1 Sampling details for all *Zosterisessor ophiocephalus* samples indicating location, sampling date, number of individuals assayed at gene–enzyme systems (N_{g-e}) and at microsatellite loci (N_{ms}), and class of ecological risk (ERC) as reported by Critto and Marcomini (2001)

Station	Geographic coordinates	ERC	Sampling date	N_{g-e}	N_{ms}
1. Ca′ Zane	45°31′27″ N–12°28′47″ E	3–4	Nov 2001	9	9
			Apr 2002	16	
			May 2002	34	18
2. Ponte della Libertà	45°27′31″ N–12°16′44″ E	6–7	Mar 2002	71	24
			Sep 2002	17	
3. Lago dei Teneri	45°24′07″ N–12°13′11″ E	6–7	Oct 2001	27	27
			Nov 2001	20	9
			Mar 2002	27	
			Apr 2002	23	18
4. Lido	45°24′26″ N–12°19′59″ E	3–4	Oct 2001	18	18
			Nov 2001	30	7
			Apr 2002	30	
			May 2002	20	18
5. Chioggia	45°13′40″ N–12°14′37″ E	2–3	Nov 2001	44	26
			Mar 2002	12	
			Apr 2002	14	
			May 2002	22	18

influence of a marine inlet (San Felice Channel), and is representative of the Northern basin. The bottom is muddy, patchy and covered with macroalgae (mainly *Ulva* sp.). This station shows a moderate level of ecological risk (ERC 3–4). Station Ponte della Libertà is located in an inner area of the Central basin, near the Marghera

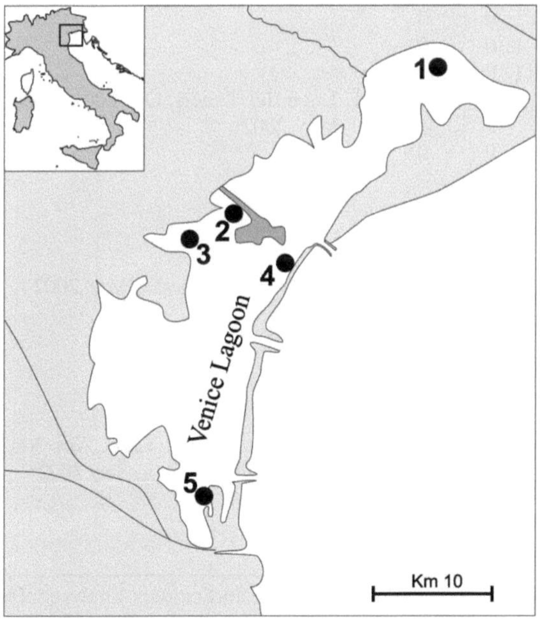

Fig. 1 Sampling locations of *Zosterisessor ophiocephalus*, labelled as in Table 1

industrial area, and shows a higher level of pollution (ERC 6–7). The same ecological risk class is attributed to the third station, Lago dei Teneri, the innermost of the Central basin, within a large tidal marsh with a soft bottom covered by macroalgae. The station Lido is also located in the Central basin, near the Lido Island, which separates the lagoon from the sea. In this case, the bottom is rich of eelgrass meadow (mainly *Zostera marina*) and seasonally covered by *Ulva* sp. beds, and ecological risk class is moderate (ERC 3–4). The fifth station, located in the Southern basin near Chioggia, where the pollution is the lowest (ERC 2–3), was chosen as a control station.

After collection, the fish were brought to the laboratory in oxygenated water, anaesthetized with MS-222 (3-aminobenzoic acid ethyl ester) dissolved in the water and dissected after severing their spinal cord. All the specimens were sexed by observing the dimorphic urogenital papilla (Miller, 1984) and were measured for total length (TL ± 0.1 cm). Four or more fragments of skeletal muscle, liver, gills and both eyes were placed separately in Eppendorf tubes. The tubes were transported to the Padua laboratory with dry-ice packs and stored at –80°C until enzyme electrophoresis. In addition, one fragment of muscle was taken from the fin for DNA extraction, and stored in absolute ethanol.

Gene–enzyme systems analysis

Tissue samples (about 10 mg) were mechanically homogenized in Eppendorf tubes in 200 μl of 0.02 M Tris/HCl pH 8 and centrifuged at 15,000 rpm for 15 min in refrigerated conditions. The supernatants from each homogenate were aliquoted and stored at –40°C for comparative analyses.

Two electrophoresis techniques were used: native, carried out on cellulose acetate (SartophorSystem-Sartorius) and isoelectrofocusing, carried out on acrilamyde gels with a mixture of carrier ampholytes (Pharmalyte, PhastSystem, Amersham-Pharmacia). The best experimental conditions were reported in Bisol (2002) and available on request. The enzyme systems analyzed are reported in Table 2. Nomenclature for protein-coding loci followed the recommendations of Shaklee et al. (1990). The proportion of polymorphic loci in each sample, and the level of observed and expected heterozygosity were calculated by Fstat software (Goudet, 2002). Heterogeneity of allele frequency distributions was analysed using a Monte Carlo simulation approach with 10,000 randomizations to account for small sample sizes and empty cells in the contingency matrix (Roff & Bentzen, 1989).

Microsatellite analysis

Total genomic DNA was extracted using the DNAeasy Blood and Tissue Extraction kit (QIAGEN) following manufacturer instructions. Genomic DNA was quantified and diluted to a working concentration of 50 ng μl^{-1}.

Eleven polymorphic microsatellite loci specific for *Zosterisessor ophiocephalus* were used for genetic analysis (Table 3); primer sequences and polymerase chain reaction (PCR) conditions are reported in Gallini et al. (2005). The forward primer for each pair was labelled with a fluorescent dye (HEX, 6-FAM, TAMRA), and PCR products were run on ABI Prism 3100 or 3700 automated sequencers. Sizing was obtained by comparison with the internal standard GS 400 Hd

Table 2 *Zosterisessor ophiocephalus* gene–enzyme variation

Enzyme (E.C. no.)	Locus	Tissue	N	n	H_{obs}	H_{exp}	Heterozygotes
Adenilate kinase (2.7.4.3)	AK*	M	434	1	0	0	
Fumarate hydratase (3.2.1.2)	FH*	M,L	434	1	0	0	
Glucose-6-phosphate isomerase (5.3.1.9)	GPI-A*	All	434	1	0	0	
	GPI-B*	All	434	2	0.005	0.005	1, Lago dei Teneri, Oct 2001 1, Lido, May 2002
Glyceraldehyde-3-phosphate dehydrogenase (1.2.1.12)	GAPDH*	All	434	1	0	0	
Isocitrate dehydrogenase NADP+ (1.1.1.42)	ICD-1*	L	434	1	0	0	
	ICD-2*	L,M,E	434	1	0	0	
L-Lactate dehydrogenase (1.1.1.27)	LDH-A*	All	434	1	0	0	
	LDH-B*	All	434	2	0.002	0.002	1, Ponte della Libertà, Mar 2002
	LDH-C*	Eye	434	1	0	0	
Malate dehydrogenase (1.1.1.37)	sMDH-1*	All	434	1	0	0	
	sMDH-2*	All	434	1	0	0	
	mMDH*	All	434	1	0	0	
Malic enzyme NADP+ (1.1.1.40)	sMEP*	All	434	1	0	0	
Phosphoglucomutase (5.4.2.2)	PGM-A*	M,L,E	434	2	0.005	0.005	1, Lido, Oct 2001 1, Lido, Apr 2002
	PGM-B*	M,L	434	2	0.002	0.002	1, Ponte della Libertà, Sep 2002
Phosphogluconate dehydrogenase (1. 1.1.44)	PGDH*	All	434	1	0	0	
Superoxide dismutase (1.15.1.1)	SOD*	L	434	1	0	0	

Reported are: enzyme name and E.C. number, locus name, tissue of expression, number of individuals analysed (N), number of alleles (n), observed heterozygosity (H_{obs}), expected heterozygosity (H_{exp}), When more than one allele was found, the number of heterozygotes and the sample of origin is also indicated

M = muscle, L = liver, E = eye

Table 3 *Zosterisessor ophiocephalus* microsatellites

Locus	N	n	H_{obs}	H_{exp}	P
ZoAC3	192	3	0.016	0.015	1
ZoAC8	192	1	0	0	–
Zo++2	192	1	0	0	–
Zo++3	190	7	0.147	0.139	1
Zo++2m 13	192	4	0.042	0.041	1
Zo++6m 13	192	2[a]	0	0	–
Zo++16m 13	192	13	0.510	0.536	0.66
Zo++22m 13	184	7	0.707	0.708	0.08
Zo++26m 13	192	12	0.833	0.801	0.39
Zo++32m 13	192	10	0.562	0.559	0.81
Zo++37m 13	191	14	0.482	0.457	0.73

Reported are: locus name, number of individuals analysed (N), number of alleles (n), observed heterozygosity (H_{obs}), expected heterozygosity (H_{exp}), and the uncorrected probability of Hardy–Weinberg equilibrium (P)

[a] Overall frequency of the commonest allele is above 99%

Rox (Applied Biosystem). Scoring was performed using the program Genotyper 3.7 (Applied Biosystem).

Number of alleles, heterozygosity, departure from Hardy–Weinberg (HW) equilibrium, and linkage disequilibrium were calculated using GENEPOP ver3.1b (Raymond & Rousset, 1995).

Pattern of population differentiation was investigated by analysis of molecular variance (AMOVA, Excoffier et al., 1992) using ARLEQUIN Ver. 2000 (Schneider et al., 2000; available at http://anthropologie.unige.ch/arlequin/). The percentage of molecular variation attributable to sample subdivision (ϕ_{st}), was calculated subdividing individuals by collection, year, sampling station, ecological risk class and sex. In addition, subdivision by size was performed for males, because in this sex there is a close relationship between age and size. Thresholds for size differences were chosen up to 9.9 cm for small and starting from 13.4 cm for big males; individuals between these two values were excluded from the analysis. This length subdivision reflects, in the Venice Lagoon population, differences between the one-year old cohort and older fish (Mazzoldi et al., 2000). Comparisons were made between small fish collected in 2002 and big fish collected in 2001, thus testing differences between the "adult" reproducing male population of 2001 and males recruiting in 2002. In all cases, statistical significance was tested by multiple permutations (10,000) of the original data set.

Results

Gene–enzyme systems

The native and IEF profiles showed multiple forms and differential tissue expression for most enzymes (Bisol, 2002), a feature common in teleost fish (Basaglia, 2002). The interpretation of electrophoretic patterns was achieved by application of the Mendelian law of monohybrid segregation and the use of statistical rules to combine the enzyme subunits. All data were compared with the previous descriptions reported in the literature concerning *Z. ophiocephalus* (Callegarini & Ricci, 1973; McKay & Miller, 1991; Miller et al., 1994; Sorice & Caputo, 1999), other Gobiidae (Wallis & Beardmore, 1984a, b; Pezold & Grady, 1989; Aizawa et al., 1994) and other fish (Basaglia, 1989; Farias et al., 1997). In this way, 18 loci were postulated for the 12 enzyme systems assayed (Table 2).

For 14 loci, only one allele was found in the 434 individuals analysed. In the other 4 loci (*GPI-B**, *LDH-B**, *PGM-A**, *PGM-B**) two alleles were scored, with one predominant allele and a second, very uncommon one (Table 2). Using the $P_{0.99}$ criteria, all loci were monomorphic at the whole sample level. *GPI-B** was polymorphic in the samples collected from station Lago Teneri in October 2001 and Lido in May 2002, *PGM-A** was polymorphic in two samples from station Lido (October 2001 and April 2002), and *PGM-B** was variable only in the sample collected from station Ponte della Libertà in September 2002. Two alleles were scored at the *LDH-B** locus in the sample from this area in March 2002, but the locus could be not classified as polymorphic because the frequency of the most common allele was 0.993.

Given the very low enzymatic variability, heterozygosity estimates per locus were found to be close or equal to 0, with an average multilocus value of 0.0008. No significant deviation from Hardy–Weinberg expectations was observed.

The lack of genetic variability prevented meaningful comparisons and no significant heterogeneity of allelic distribution was observed between males and females, between sites or season of collection, between sites pooled by class of ecological risk, and between polymorphic (Lido, Lago dei Teneri and Ponte della Libertà) and monomorphic (Ca′ Zane and Chioggia) samples.

Microsatellites

Microsatellite analysis of 192 individuals with 11 microsatellite loci revealed substantial molecular variation (Table 3). Eight loci were found to be polymorphic (99% criterium) with 3–14 alleles per locus. Some loci turned out to be highly polymorphic (Zo++26m13, Zo++22m13) with the frequency of the most common allele that ranged from 0.33 to 0.39. All loci were found to be in Hardy–Weinberg equilibrium. One locus (Zo++22m13) showed a *P*-value close to the uncorrected threshold of 0.05 ($P = 0.08$), but a Fisher test did not reject the hypothesis of global HW equilibrium across loci ($P = 0.55$). No evidence of linkage disequilibrium between loci was found.

Table 4 reports the level of variability for different samples. No significant deviation from Hardy–Weinberg equilibrium was detected. Molecular variability was found to be quite different between sites, with the observed heterozygosity ranging from 0.254 in the sample collected at Ca′ Zane in 2002 to 0.332 for the sample from Lago dei Teneri 2002; differences were not statistically significant.

ϕ_{st} statistics of differentiation were estimated following different hypotheses. First, we tested the hypothesis that collections of individuals represented samples taken from a single panmictic population. Accordingly, we calculated ϕ_{st} for 9 samples of individuals collected in the same year in a given sampling station (Table 1). Results showed positive, but not statistically significant, values of ϕ_{st} ($\phi_{st} = 0.00037$, $P = 0.36$) indicating overall genetic homogeneity in the Venice Lagoon. Comparisons between five samples representative of the site of collection provided negative values of ϕ_{st} indicating the lack of stable local differentiation ($\phi_{st} = -0.00060$, $P = 0.52$). Similar results were obtained in comparisons between animals collected in different years ($\phi_{st} = -0.00054$, $P = 0.54$), by comparison between the two sexes ($\phi_{st} = -0.00149$, $P = 0.77$), and by comparisons between areas with different class of ecological risk ($\phi_{st} = -0.00336$, $P = 0.88$).

However, when males were subdivided by size, the comparison between small individuals collected in 2002 ($N = 21$) and large individuals collected in 2001 ($N = 55$), provided a positive and statistically significant value of ϕ_{st}

Table 4 *Zosterisessor ophiocephalus* heterozygosity

Sample	N	H_{obs}	H_{exp}	P
Chioggia 2001	26	0.305 (0.311)	0.322 (0.336)	0.383
Chioggia 2002	18	0.298 (0.346)	0.289 (0.320)	0.297
Ca′ Zane 2001	9	0.293 (0.363)	0.269 (0.320)	0.973
Ca′ Zane 2002	18	0.254 (0.289)	0.266 (0.313)	0.739
Lido 2001	25	0.320 (0.328)	0.307 (0.307)	0.838
Lido 2002	18	0.271 (0.295)	0.289 (0.323)	0.927
Lago dei Teneri 2001	36	0.298 (0.343)	0.289 (0.323)	0.397
Lago dei Teneri 2002	18	0.332 (0.367)	0.307 (0.317)	0.999
Ponte della Libertà 2002	24	0.294 (0.301)	0.299 (0.315)	0.944

(ϕ_{st} = 0.0107, P = 0.05). The threshold used was chosen to separate younger (one year old) fish recruiting in 2002 from older males reproducing in 2001, based on published data that indicate a close relationship between age and length in this sex (Mazzoldi et al., 2000). The observed differentiation was robust against slight changes of the size threshold, and indicates significant differences between age classes. The average expected heterozygosity was found to be lower in small males (0.286 ± 0.178) than in big ones (0.370 ± 0.212), but the difference was not statistically significant.

Discussion

Our study shows an extremely low level of genetic variation at gene–enzyme systems of *Z. ophiocephalus* from Venice Lagoon. Previous studies, carried out with a similar set of enzymes on very few specimens of this species reported different levels of average heterozygosity. Values ranged from 0.08 in 10 individuals from Grado Lagoon (about 70 km North of Venice Lagoon; Miller et al., 1994) to 0.00 in 20 specimens collected off the coast of Ancona (about 200 km South of Venice Lagoon; Sorice & Caputo, 1999), and was 0.04 in 5 individuals from the Venice Lagoon (McKay & Miller, 1991). Our value of 0.0008 supports the result of Sorice & Caputo (1999), and is interesting when compared with the average value of 0.051 calculated for 195 species of teleosts (Ward et al., 1992). Similarly low levels of allozyme variation have been reported in other fish: for instance, about 7% of species showed heterozygosities close to 0 in a survey of 106 marine teleosts (Smith & Fujio, 1982); this result was explained by a major taxonomic effect and by a habitat specialist-generalist model. Interestingly, average heterozygosities reported for Perciformes (H = 0.05; Smith & Fujio, 1982) and Gobidae (H = 0.04, Miller et al., 1994; H = 0.05, Sorice & Caputo, 1999) are very close to the average value of other teleosts, indicating that the low level of genetic variation could be a peculiarity of *Z. ophiocephalus*.

The direct comparison of loci shared with previous studies reveals striking differences for the loci LDH-B* and sMEP* in the Venice Lagoon (McKay & Miller, 1991) and IDHP-2* in the Marano Lagoon (Miller et al., 1994). In fact, the three loci were formerly found to be highly heterozygous, with a frequency for the most common allele of 0.8, 0.60, and 0.75, respectively. As far as it is possible to compare results obtained with different techniques at different times, these data suggest that genetic variation has been eliminated in the Venice Lagoon population during the last few years at these three loci. In this case, the present day lack of heterozygosity could represent an example of rapid evolutionary change in response to environmental pressure. This conclusion seems reasonable, because in the present study additional efforts to detect additional alleles with different electrophoretic conditions were made without success.

In contrast with the low polymorphism of allozyme loci, microsatellite analysis showed the existence of considerable genetic variation among 192 *Z. ophiocephalus* individuals, with an average expected heterozygosity of 0.296. This result, obtained with DNA level 'neutral' markers, seems to exclude a history of a small population size for the Venice Lagoon and suggests the need for further investigations aimed at addressing the relative fitness of rare heterozygotes, and the existence of recent non-equilibrium population dynamic processes.

The latter aspect has been in part investigated in the present study by microsatellites. In this case, the existence of genetic variation allowed to test the hypothesis of genetic homogeneity between samples of grass goby collected inside the Venice Lagoon. Despite the consistent amount of polymorphism, no significant variation was detected between sites of collection, between years of collection, between sexes, or between areas with different degree of pollution, indicating that the population apparently behaves as a single panmictic unit. In contrast, when analysing individuals of different size, our results indicated significant differences between the small males collected in 2002 and the large ones collected in 2001. In turn, the size difference test reflects differences between the reproductive population of males in 2001 and the recruiting one-year old male cohort of 2002, because a strong relationship exists between age and size for males of this species (Mazzoldi et al., 2000). This result indicates a complex population dynamics for the species. Differences between age classes can be explained by a number of factors, including recruitment from genetically differentiated populations, variance in reproductive success (Hedgecock, 1994), drift linked to poor recruitment, and selective mortality. In the present study the impact of gene flow can be excluded because the only candidate population for exchanging individuals, the nearby Marano Lagoon population, resulted, during a preliminary survey, genetically homogeneous with the Venice Lagoon population (Rianna, 2003). In addition, recruitment from genetically differentiated populations should produce in many cases a higher genetic diversity in recruits, which was not observed in this case.

Our finding of statistically significant differences between age classes thus provides support for the hypothesis that present-day processes due to drift or natural selection are modifying the genetic composition of the species in the Venice Lagoon. Our results suggest that a careful monitoring plan should be developed for this species, aimed at estimating the effective population size through genetic and demographic approaches, and at estimating the adaptive significance of the residual allozyme variability.

Acknowledgements The Authors are grateful to Prof. Danilo Mainardi and Prof. Patrizia Torricelli for their hospitality in the laboratory of the University of Venice, and to Paola Campagnolo for her technical assistance. This work was supported by a Corila (Consortium for Coordination of Research Activities Concerning the Venice Lagoon System) grant.

References

Aizawa, T., M. Hatsumi & K. Wakahama, 1994. Systematic study on the *Chenogobius* species (Family Gobiidae) by analysis of allozyme polymorphisms. Zoological Science 2: 455–465.

Basaglia, F., 1989. Some aspects of isozymes of lactate dehydrogenase and glucosephosphate isomerase in fish. Comparative Biochemistry and Physiology Part B 92: 213–226.

Basaglia, F., 2002. Multilocus isozyme systems in African lungfish, *Protopterus annectens*: distribution, differential expression and variation in dipnoans. Comparative Biochemistry and Physiology Part B 131: 89–102.

Bisol, P. M., 2002. Ecogenetic biodiversity in *Zosterisessor ophiocephalus* from the lagoon of Venice. I: Gene–enzyme polymorphisms. In Campostrini, P. P. (ed.), Scientific Research and Safeguarding of Venice. Corila Research Program 2001 results. Istituto Veneto SS LL AA, Venezia, 601–608.

Callegarini, C. & D. Ricci, 1973. Lactate dehydrogenase (LDH) isozyme in some species of fresh water, euryhaline and salt water teleosts from the Po plain and its sea-coasts. Bollettino di Zoologia 40: 25–30.

Carvalho, G. R. & L. Hauser, 1998. Advances in the molecular analysis of fish population structure. Italian Journal of Zoology 65: 21–33.

Critto, A. & A. Marcomini, 2001. Rischio ecologico e inquinamento chimico lagunare. Libreria Editrice Cafoscarina, Milano.

Excoffier, L., P. Smouse & J. Quattro, 1992. Analysis of molecular variance inferred from metric distances among DNA haplotypes: application to human mitochondrial DNA restriction data. Genetics 131: 479–491.

Farias, I. P., M. N. Paula-Silva & V. M. Almeida-Val, 1997. No co-expression of LDH-C in Amazon cichlids. Comparative Biochemistry and Physiology Part B 117: 315–319.

Franco, A., S. Malavasi, F. Pranovi, F. Franzoi & P. Torricelli, 2002. Preliminary data on gonadal development and fecundity in the Grass goby, *Zosterisessor ophiocephalus* (Pallas, 1811) from the Venice Lagoon (Northern Italy). Acta Adriatica 43: 43–48.

Gallini, A., L. Zane & P. M. Bisol, 2005. Isolation and characterization of microsatellites in *Zosterisessor ophiocephalus* (Perciformes, Gobiidae). Molecular Ecology Notes 5: 24–26.

Goudet, R., 2002. Fstat, version 2.9.3.2. http://www2.unil.ch/izea/softwares/fstat.html.

Hedgecock, D., 1994. Does variance in reproductive success limit effective population sizes of marine organisms? In Beaumont, A. R. (ed.), Genetics and Evolution of Aquatic Organisms. Chapman & Hall, London, 122–134.

Hedgecock, D., V. Chow & R. S. Waples, 1992. Effective population numbers of shell-fish broodstocks estimated from temporal variance in allelic frequencies. Aquaculture 108: 215–232.

Hutchinson, W. F., C. van Oosterhout, S. I. Rogers & G. R. Carvalho, 2003. Temporal analysis of archived samples indicates marked genetic changes in declining North Sea cod (*Gadus morhua*). Proceeding Royal Society London Ser B: Bio Sci 270: 2125–2132.

Ikeda, M., M. Nunokawa & N. Taniguchi, 2003. Lack of mitochondrial gene flow between populations of the endangered amphidromous fish *Plecoglossus altivelis ryukyuensis* inhabiting Amami-oshima Island. Fishery Sciences 69: 1162–1168.

Mainardi, D., R. Fiorin, A. Franco, P. Franzoi, O. Giovanardi, A. Granzotto, A. Libertini, S. Malavasi, F. Pranovi, F. Riccato & P. Torricelli, 2002. Fish diversity in the Venice Lagoon: preliminary report. In

Campostrini, P. P. (ed.), Scientific Research and Safeguarding of Venice. Corila Research Program 2001 results. Istituto Veneto SS LL AA, Venezia, 583–594.

Maltagliati, F., P. Domenici, C. F. Fosch, P. Cossu, M. Casu & A. Castelli, 2003. Small-scale morphological and genetic differentiation in the Mediterranean killifish *Aphanius fasciatus* (Cyprinodontidae) from a coastal brackish-water pond and an adjacent pool in northern Sardinia. Oceanological Acta 26: 111–119.

Marconato, A., M. B. Rasotto & C. Mazzoldi, 1996. On the mechanism of sperm release in three gobiid fishes (Teleostei: Gobiidae). Environmental Biology of Fishes 46: 321–327.

Mazzoldi, C., M. Scaggiante, E. Ambrosin & M. B. Rasotto, 2000. Mating system and alternative male mating tactics in the grass goby *Zosterisessor ophiocephalus* (Teleostei: Gobiidae). Marine Biology 137: 1041–1048.

McKay, S. I. & P. J. Miller, 1991. Isozyme criteria in the testing of phyletic relationships between species of *Gobius* and related Eastern Atlantic-Mediterranean genera (Teleostei: Gobiidae). Journal of Fish Biology 39: 291–299.

Miller, P. J., 1984. The tokology of Gobioid fishes. In Potts, G. W. & R. J. Wootton (eds), Fish Reproduction: Strategies and Tactics. Academic Press, London, 119–153.

Miller, P. J., M. Serventi, D. Soregaroli, P. Torricelli & G. Gandolfi, 1994. Isozyme genetics and the phylogeny of Italian freshwater gobies (Teleostei: Gobioidei). Journal of Fish Biology 44: 439–451.

Ota, D., M. Marchesan & E. A. Ferrero, 1996. Sperm release behaviour and fertilization in the grass goby. Journal of Fish Biology 49: 246–256.

Penzo, E., G. Gandolfi, L. Bargelloni, L. Colombo & T. Patarnello, 1998. Messinian salinity crisis and the origin of freshwater lifestyle in Western Mediterranean gobies. Molecular Biology and Evolution 15: 1472–1480.

Pezold, F. & J. M. Grady, 1989. A morphological and allozymic analysis of species in the *Gobionellus oceanicus* complex (Pisces:Gobiidae). Bulletin of Marine Science 45: 648–663.

Planes, S, M. Parroni & C. Chauvet, 1998. Evidence of limited gene flow in three species of coral reef fishes in the lagoon of New Caledonia. Marine Biology 130: 361–368.

Raymond, M. & F. Rousset, 1995. Genepop version 1.2: population genetics software for exact tests and ecumenism. The Journal of Heredity 86: 248–249.

Rianna, E., 2003. Biodiversità in *Zosterisessor ophiocephalus* di due lagune dell'Alto Adriatico. Bachelor thesis (in Italian), University of Padova.

Roff, D. A. & P. Bentzen, 1989. The statistical analysis of mitochondrial DNA polymorphisms: χ^2 and the problem of small samples. Molecular Biology and Evolution 6: 539–545.

Ruzzante, D. E., C. T. Taggart & D. Cook, 1996. Spatial and temporal variation in the genetic composition of a larval cod (*Gadus morhua*) aggregation: cohort contribution and genetic stability. Canadian Journal Fish Aquatic Science 53: 2695–2705.

Scaggiante, M., C. Mazzoldi, C. W. Petersen & M. B. Rasotto, 1999. Sperm competition and mode of fertilization in the grass goby *Zosterisessor ophiocephalus* (Teleostei: Gobiidae). The Journal of Experimental Zoology 283: 81–90.

Schneider, S., D. Roessli & L. Excoffier, 2000. Arlequin ver. 2.000. A Software for Population Genetic Data Analysis. Genetics and Biometry Laboratory, University of Geneva, Switzerland.

Shaklee, J. B., F. W. Allendorf, D. C. Morizot & G. S. Whitt, 1990. Gene nomenclature for protein-coding loci in fish. Transactions of the American Fisheries Society 119: 2–15.

Shaklee, J. B. & P. Bentzen, 1998. Genetic identification of stocks of marine fish and shellfish. Bulletin of Marine Science 62: 589–621.

Smith, P. J. & Y. Fujio, 1982. Genetic variation in marine teleosts: high variability in habitat specialists and low variability in habitat generalists. Marine Biology 69: 7–20.

Sorice, M. & V. Caputo, 1999. Genetic variation in seven goby species (Perciformes: Gobiidae) assessed by electrophoresis and taxonomic inference. Marine Biology 134: 327–333.

Taylor, M. S. & M. E. Hellberg, 2003. Genetic evidence for local retention of pelagic larvae in a Caribbean reef fish. Science 299: 107–109.

Torricelli, P., S. Malavasi, N. Novarini, F. Pranovi & D. Mainardi, 2000. Elongation of fin rays in parental males of *Zosterisessor ophiocephalus* (Pisces, Gobiidae). Environmental Biology of Fishes 58: 105–108.

Wallis, G. P. & J. A. Beardmore, 1984a. An electrophoretic study of the systematic relationships of some closely related goby species (Pisces, Gobiidae). Biological Journal of Linnean Society 22: 107–123.

Wallis, G. P. & J. A. Beardmore, 1984b. Genetic variation and environmental heterogeneity in some closely related goby species. Genetica 62: 223–237.

Ward, R. D., D. O. F. Skibinsky & M. Woodwark, 1992. Protein heterozygosity, protein structure, and taxonomic differentiation. Evolutionary Biology 26: 73–157.

Ward, R. D., M. Woodward & D. O. F. Skibinsky, 1994. A comparison of genetic diversity levels in marine, freshwater, and anadromous fishes. Journal of Fish Biology 44: 213–232.

Hydrobiologia (2007) 577:161–168
DOI 10.1007/s10750-006-0425-4

LAGOONS AND COASTAL WETLANDS

Ecological engineering in intertidal saltmarshes

James T. Morris

Abstract Feedbacks between plant biomass density and sedimentation maintain intertidal marshes in equilibrium with mean sea level (MSL). Stable marshes exist at an elevation that is supraoptimal for the biomass density of marsh macrophytes. At this elevation, biomass density is sensitive to changes in MSL, and adjustments in productivity and sedimentation rate help to maintain the marsh in a dynamic equilibrium with sea level, provided that the surface elevation remains within the supraoptimal range of the vegetation. The equilibrium elevation varies inversely with the rate of sea-level rise and directly with biomass density. It was also shown that a succession of intertidal plant communities depends upon the rate of sea level rise and the distribution of biomass density as a function of hydroperiod. Soft engineering solutions to coastal flooding could incorporate planting of marsh vegetation in the intertidal zone for the purpose of promoting sedimentation and dissipating wave energy. A successful design would employ plant species that have varying degrees of tolerance to flooding, maximum drag at their preferred depths, broad ranges within the intertidal zone, and that form a successional series.

Keywords Spartina · Saltmarsh · Sediment · Accretion · Ecological engineering · Sea level

Introduction

There is considerable interest in the response of coastal wetlands to rising sea level due to the ecological and commercial importance of these ecosystems and because of an anticipated acceleration in the rate of sea-level rise due to global warming. Salt marshes are known to have maintained an elevation in equilibrium with sea level during the late Holocene by accumulating mineral sediment and organic matter at a rate that compensates for rising sea level (Redfield, 1972). Previous models predict that intertidal marshes approach an equilibrium elevation that approximates that of mean high water (Krone, 1985). This is based on logic that posits that the rate of sedimentation is proportional to the depth of water overlying the marsh at high tide. As this depth approaches zero at mean high water (MHW), net sedimentation rate also approaches zero. If flood depth increases, all else being equal,

Guest editors: P. Viaroli, P. Lasserre & P. Campostrini
Lagoons and Coastal Wetlands in the Global Change Context: Impacts and Management Issues

J. T. Morris (✉)
Department of Biological Sciences and Belle W. Baruch Institute for Marine and Coastal Sciences, University of South, Columbia, South Carolina 22908, USA
e-mail: morris@biol.sc.edu

then the quantity of sediment suspended above the surface and settling time will increase. This should result temporarily in a higher rate of sedimentation until the flood depth decreases and a new equilibrium is reached. However, this simple model ignores important feedbacks between the marsh vegetation, hydroperiod, and sedimentation rate.

Recent work in a North Inlet marsh, South Carolina, USA, has shown that the relative elevation of the sediment surface is a critically important variable that ultimately controls the productivity of the saltmarsh plant community and that productivity has a positive feedback on the rate of accretion of the marsh surface (Morris et al., 2002). From the perspective of protecting infrastructure along crowded coastlines, in addition to the positive benefits of accreting sediment, vegetated marshes effectively dissipate wave energy (Moller et al., 1999; Danielsen et al., 2005). These are species-specific characteristics that can be modified or engineered by the appropriate selection of species or cultivars. This paper presents a strategy for marsh restoration and reports results of a modelling study designed to explore the potential for vegetated marshes to raise the elevation of their habitat under different sea-level rise scenarios and to illustrate the effect of inducing a succession of marsh communities or species with different ranges of tolerance to flooding for maximum gain in elevation.

Materials and methods

The fieldwork upon which this report is based was conducted within North Inlet estuary, where a long-term study of salt marsh primary production has been in place since 1984 (Morris, 2000). North Inlet is a 32 km^2, tidally dominated saltmarsh estuary with a watershed area of about 75 km^2 and minimal freshwater input. Approximately 82% of the estuary is intertidal salt marsh, dominated by the grass *Spartina alterniflora* Loisel., and mud flat, and 18% is open water. North Inlet experiences a regular semi-diurnal tidal pattern with a mean tidal range of about 1.4 m (Finley, 1975).

We began a parallel study of marsh accretion in 1996. This study was conducted on replicate (n = 3) control and nutrient-treated plots that were dominated by a monoculture of short *S. alterniflora* in the middle of the marsh platform. The elevation of the site is about 45 cm above mean sea level (Morris et al., 2005). Elevation change was measured with a Surface-Elevation Table (SET), which is a portable, mechanical leveling device designed to attach to a stable, benchmark pipe driven into the ground (Boumans & Day, 1993; Cahoon et al., 2002). This experiment and published data on sediment accretion based on ^{210}Pb (Vogel et al., 1996) provided the data needed to estimate coefficient values used in the model reported here. Details of the design of the field experiments are given in Morris et al. (2002).

Results

Model description

There exist both upper and lower limits of relative elevation for the biomass density (B) of *S. alterniflora* and other intertidal macrophyte species. The lower limits are likely to be functions of the duration of flooding and hypoxia, while the upper elevations are determined by salt stress, dessication, or competitive pressure from other species (Bertness, 1991; Emery et al., 2001; Bertness & Ewanchuk, 2002; Pennings et al., 2005). The biomass distribution of a given species or plant community is assumed here to have a shape resembling a parabola. That is, a species is assumed to have a maximum biomass when growing at an optimum elevation (relative to mean sea level), and its biomass is diminished at elevations near the limits of its growth range. These curves can be viewed as dimensions of a species' fundamental (in the absence of competitors) or realized (in the presence of competitors) niche, sensu Hutchinson (1957). Different plant species or communities will have different optima and ranges that may or may not overlap. These distributions are defined by the coefficients a_i, b_i and c_i in Eq. 1, where subscript i refers to a specific dominant plant species or community type. These curves are assumed to be invariant for simulations reported here, but in reality the shapes of the curves may change depending on competitive interactions between species at the boundaries of their distributions or changes in the physical or chemical environment.

$$B_i = a_i D + b_i D^2 + c_i \tag{1}$$

The values of the coefficients a, b, and c will also differ regionally as a function of tidal range, salinity, or climate. Variable D is the depth of the marsh surface below MHW. The species distributions relative to D will vary regionally with tidal amplitude and probably climate. As tidal amplitude increases, the vertical range of a given intertidal species will also increase (McKee & Patrick, 1988).

We showed previously (Morris et al., 2002) that the rate of change of elevation of the marsh surface ($\mathrm{d}Y/\mathrm{d}t$) is a positive function of the standing density of aboveground biomass (B). For simplicity this relationship is assumed to be linear over a limited range of depth $D \geq 0$ below mean high water:

$$\mathrm{d}Y/\mathrm{d}t = (q + \sum k_i B_i)D \tag{2}$$

where q and k_i are parameters that are proportional to the rate of sediment loading (q) and the efficiency of the vegetation (k_i) as a sediment trap. The values of q and k are likely to vary locally, seasonally, and regionally as a function of sediment availability and tidal range (Stevenson et al., 1986; Temmerman et al., 2003). Parameter k is analogous to a drag coefficient and will differ among species (Leonard & Luther, 1995). Parameter q represents the net effects of sedimentation and erosion and, therefore, will vary spatially within a marsh, but is assumed here to be a constant. Thus, Eq. 2 as used here is zero dimensional, i.e., it represents the change in elevation of a single point within the marsh.

The contribution of the vegetation to the sedimentation rate is given by $\sum k_i B_i D$, but B_i is not a constant as it varies with D as given by Eq. 1, Eq. 2 may be rewritten as follows, after substituting for B from Eq. 1:

$$\mathrm{d}Y/\mathrm{d}t = \sum k_i b_i D^3 + \sum k_i a_i D^2 + (q + \sum k_i c_i)D \tag{3}$$

The single species solution

In a marsh dominated by *Spartina alterniflora* in North Inlet estuary, model coefficients q and k were estimated to be 0.0018 and 1.5×10^{-5}, respectively, based on ^{210}Pb estimates of sediment accretion of 0.27 cm y^{-1} (Vogel et al., 1996) and on differences in recent rates of sediment accretion between control and fertilized plots (Morris et al., 2002). Coefficient k as used here operationally accounts for the contribution to vertical accretion due to the production and accumulation of organic matter, because of the method used to calibrate the model. Furthermore, coefficient k was calibrated to the rate of annual primary production. However, the contribution to sedimentation of suspended particles by plants depends directly on the biomass density, not the primary production. Although production and biomass density are related, it would be preferable to define k in terms of biomass density. Based on long-term measurements in North Inlet marshes it was concluded that the relationship between maximum biomass density and annual production is site-specific, but a reasonable P/B ratio for the short form of *S. alterniflora* here is about 1.6 (Morris in press). The tall form of *S. alterniflora* has a lower ratio, 1.2, but the short form dominates the marsh platform where the model was calibrated. Using this ratio the k coefficient is recomputed as 2.4×10^{-5} cm g^{-1} y^{-1}, which is the value used in simulations here, and variable B is expressed as the maximum biomass (g m^{-2}).

The equilibrium depth (D) can be solved after substituting different values for the accretion rate ($\mathrm{d}Y/\mathrm{d}t$), and the accretion rate can be decomposed into its different components by solving Eq. 3 after alternately zeroing parameters q and k. The result demonstrates that the plants in this marsh are largely responsible for trapping sediment and promoting growth in vertical elevation (Fig. 1). The line labeled C is the accretion rate in

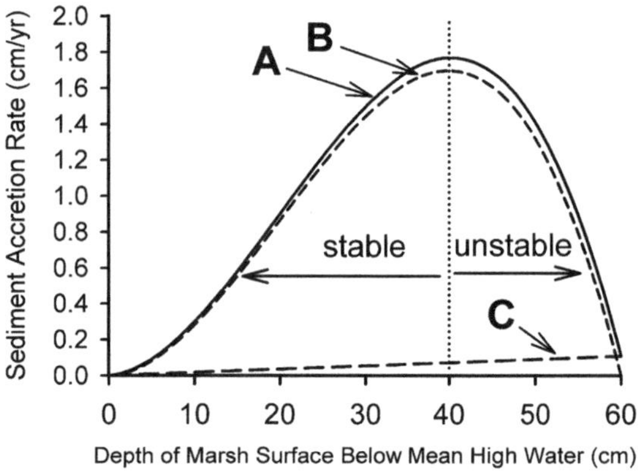

Fig. 1 Solutions to Eq. 3 showing the relative contribution of plants to sediment accretion. In C the plants were removed by zeroing the parameter *k*. Curve B represents the contribution of only the vegetation (parameter *q* was zeroed) and curve A represents the sum of curves B and C. Only solutions for depths less than 40 cm are feasible (see text)

the absence of plants, while line *B* is the contribution made by plants (Fig. 1). In the presence of a dense canopy of marsh grass, the accretion rate greatly exceeds that of a mud flat. Moreover, the accretion rate is predicted to be greatest at depths corresponding approximately to the maximum biomass density (Fig. 2), and accretion rate falls off rapidly as biomass approaches zero at the limits (0 and 60 cm below mean high water).

After computing equilibrium depth *D* by substituting a range of values for d*Y*/d*t* in Eq. 3, the resulting values of *D* were substituted into

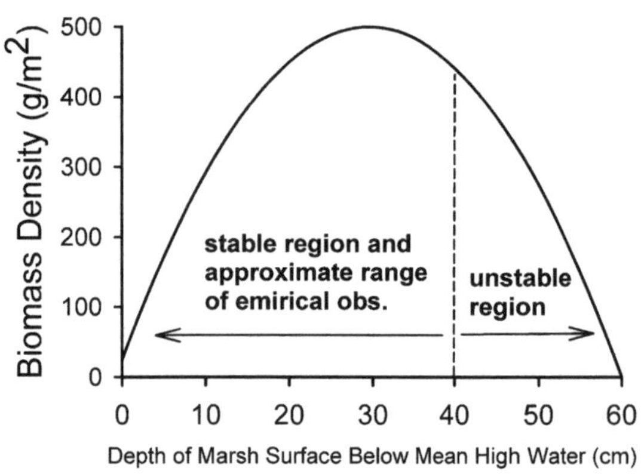

Fig. 2 Solution to Eq. 1 showing the hypothetical distribution of biomass as a function of depth below mean high water

Eq. 1 to calculate the equilibrium biomass density *B* corresponding to *D* (Fig. 2). This biomass distribution generally describes the distribution of *S. alterniflora* across the marshes in North Inlet estuary, although the symmetry of the curve, which is dictated by the use of a parabola (Eq. 1), is probably unrealistic. In theory, the vegetated marsh here is unstable at depths greater than 40 cm below MHW (Fig. 1). At depths greater than 40 cm, an increase in sea level would increase the depth, which leads to a decline in sedimentation rate (Fig. 1), biomass (Fig. 2) and, ultimately, to the loss of vegetation.

The multiple species solution

The sensitivity of the marsh-sea level interaction to the biomass distribution was demonstrated in a series of numerical simulations in which three hypothetical species were distributed across the marsh in different configurations (Figs. 3–7). The model was solved first for three species with overlapping distributions as shown in Fig. 3A in an estuary having a tidal amplitude of 60 cm and a rate of sea-level rise of 0.2 cm/y. Here (Fig. 3) the species' ranges are plotted as elevations relative to mean sea level (MSL), because relative elevation is a more intuitive benchmark than depth below MHW. Maximum biomass for each species was specified as 300 g m^{-2}, and the growth ranges collectively spanned 10–70 cm above MSL. The species had optimum depths of 25, 40 and 55 cm above MSL (Fig. 3). The marsh was given an arbitrary elevation at time zero of 25 cm above MSL, and the marsh equilibrated eventually at 27 cm above MSL, which is close to the optimum elevation for the first species. At such a low biomass the accretion of sediment was just sufficient to add 0.2 cm/y of elevation when the relative elevation was only 27 cm. Thus, the species with the lowest growth range (15–40 cm above MSL) dominated. The relative marsh elevation never increased to a level that would support either of the other two species.

The marsh response changed significantly when the maximum biomass of the first species was raised from 300 to 600 g m^{-2} (Fig. 4). Although the vertical ranges of the three species were not changed, the increase in biomass density of the

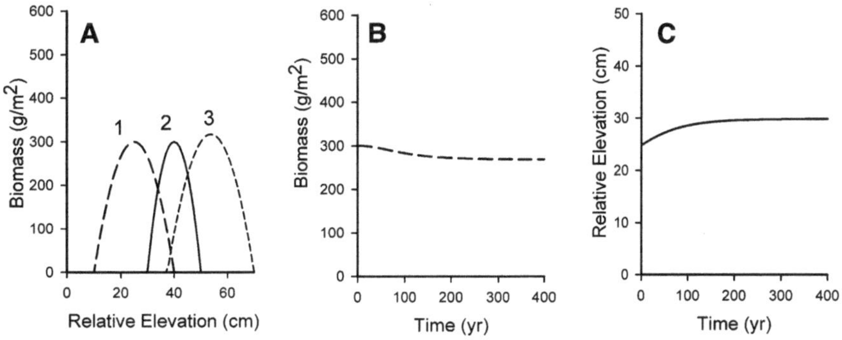

Fig. 3 Model results with three species, distributed as above (**A**), and a rate of sea-level rise of 0.2 cm/y. Panel (**B**) shows the predicted biomass trajectory as the marsh surface equilibrates at a relative elevation of 30 cm (**C**) above mean sea level (MSL)

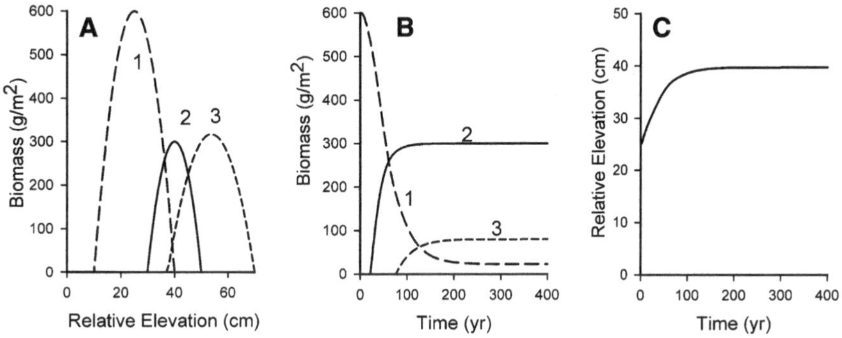

Fig. 4 Model results with three species, distributed as above (**A**), and a rate of sea-level rise of 0.2 cm/y. Panel (**B**) shows the predicted biomass trajectory of the three species as the marsh surface equilibrates at a relative elevation of 40 cm (**C**) above mean sea level (MSL)

first species was sufficient to increase the sedimentation rate and, consequently, the relative elevation. Relative elevation eventually stabilized near the optimum level of the second species or 40 cm. What occurred was a succession from a marsh dominated by one species to another as the relative elevation of the marsh surface increased. Moreover, this succession to species 2 was initiated as a consequence of the high biomass of species 1. This hypothetical marsh essentially gained 15 cm on sea level over the course of a century (Fig. 4) or about 0.15 cm y^{-1} greater than the rate of sea-levèl rise until reaching the equilibrium relative elevation of 40 cm.

In the next simulation the maximum biomass densities of both the first and second species were raised to 600 g m^{-2} (Fig. 5). This change in the biomass distribution resulted in a further increase in relative elevation to 46 cm above MSL. Although the first species disappeared completely from the species assemblage and the biomass density of the third species increased, the second species still dominated. Only by increasing the maximum biomass density of the third species was it possible to raise the equilibrium elevation to a level, 49 cm, at which the third species dominated (Fig. 6). Thus, the sequence of species successions is dependent on the biomass distributions of the individual species, although distur-

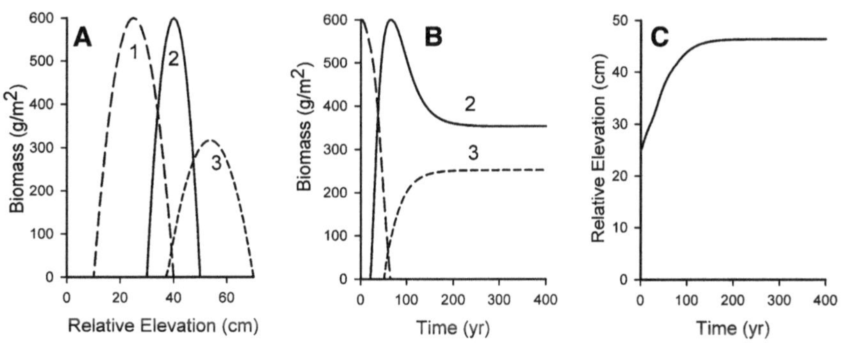

Fig. 5 Model results with three species, distributed as above (**A**), and a rate of sea-level rise of 0.2 cm/y. Panel (**B**) shows the predicted biomass trajectory of the three species as the marsh surface equilibrates at a relative elevation of 46 cm (**C**) above mean sea level (MSL)

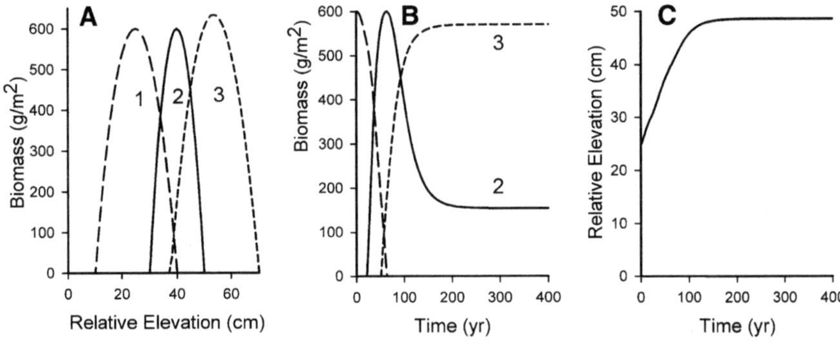

Fig. 6 Model results with three species, distributed as above (**A**), and a rate of sea-level rise of 0.2 cm/y. Panel (**B**) shows the predicted biomass trajectory of the three species as the marsh surface equilibrates at a relative elevation of 49 cm (**C**) above mean sea level (MSL)

bance events may be necessary to initiate the process of succession (Bertness & Ellison, 1987).

Succession and equilibrium elevation also depend on the rate of sea-level rise. When the biomass distributions were held constant as in the previous (Fig. 6) example, and the rate of sea-level rise was increased from 0.2 to 0.4 cm y^{-1}, the succession was halted. The relative elevation equilibrated at 30 cm above MSL (Fig. 7C). The relative elevation of the marsh surface declines with increasing rate of sea-level rise (Morris et al., 2002). Hence, species with greater tolerance for flooding will prevail as the rate of sea-level rise increases and as equilibrium elevation decreases.

Discussion

Model coefficients q and k describe the relative importance of mineral input, in the absence of vegetation, and biogenic accretion, respectively. Biogenic accretion is meant to include mineral sediment that is trapped as a consequence of standing biomass as well as organic matter accretion. As discussed earlier, the model was parameterized using data from marshes in North Inlet estuary, where the input of sediment is probably low in comparison to marshes at the mouths of major river systems. Consequently, the solution is dominated by coefficient k and plant biomass density is a major determinant of vertical sediment accretion (Fig. 1). However, it is conceivable that some marshes may be q-dominated, while others are k-dominated as in the case of North Inlet.

Marshes are stable against variations in sea level when their relative elevations are supraoptimal for the vegetation (left side of the curves in Figs. 1 and 2). At these high elevations the vegetation responds positively to a relative rise in sea level. That, in turn, increases the sedimentation rate. On the suboptimal side of the response curve (Fig. 2), an increase in sea level will decrease productivity and sedimentation rate, which increases the depth, leading eventually to conversion to mud flat. Thus, at depths less than 40 cm, negative feedback stabilizes the marsh as the rate of sea-level rise increases toward a threshold, while positive feedback destabilizes the marsh at depths greater than 40 cm. This specific conclusion applies to a marsh with a tidal amplitude of about 60 cm, and the same caveats

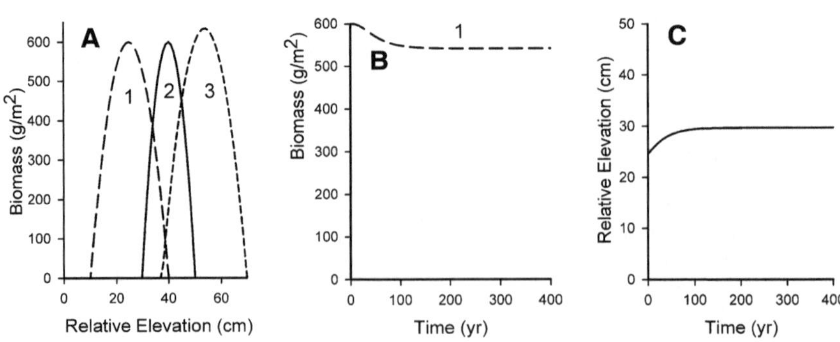

Fig. 7 Model results with three species, distributed as above (**A**), and a rate of sea-level rise of 0.4 cm/y. Panel (**B**) shows the predicted biomass trajectory as the marsh surface equilibrates at a relative elevation of 30 cm (**C**) above mean sea level (MSL)

about parameter values as discussed above apply as well. However, the general conclusion is valid that negative feedback stabilizes the marsh at supraoptimal elevations and positive feedback destabilizes at suboptimal elevations.

This depth limit of 40 cm below mean high water (Fig. 1) is consistent with an analysis of LIDAR data from North Inlet, which showed that 80% of the vegetated marsh was distributed between 14 cm and 40 cm below mean high water (Morris et al., 2005).

By definition, equilibrium depth is the relative depth that is in equilibrium with mean sea level. The equilibrium depth is given by the abscissa under the left side of curve A (Fig. 1). As the rate of sea-level rise increases, the equilibrium depth increases. An important feature of Fig. 1 is the decline in accretion rate at depths greater than 40 cm below MHW. The drop in accretion rate occurs because of a decline in the biomass density of the plant community at these depths (Fig. 2). Consequently, these depths correspond to a region of instability. Any increase in the rate of sea-level rise results in an increase in depth, and at depths greater than 40 cm an increase in depth results in a decline in primary production, a further decline in sedimentation, and conversion of the marsh to mudflat.

It is important to note that the specific response of the marsh surface to a change in sea level is highly sensitive to the biomass distribution (Fig. 2), which in turn is a function of tidal range. *S. alterniflora* occupies only the upper part of the tidal range, and as the tidal range decreases, so does the absolute range of *Spartina* (McKee & Patrick, 1988). Thus, the dynamic range of a salt marsh, or the maximum rate of sea level rise that can be tolerated, is proportional to the tidal range.

A stable marsh that is rising in equilibrium with MSL will transgress inland across the terrestrial landscape at a rate that depends on the land slope. If the marsh looses area elsewhere at a rate similar to its rate of transgression, then the total marsh area will be constant, more or less depending on the relative rates of landward transgression, change in channel volume, and barrier island migration. However, barriers to marsh transgression, such as geomorphological barriers, seawalls or other protective structures, could result in a net loss of marsh area even when the marsh interior is stable. This depends on the stability of the other internal and external marsh boundaries.

Conclusion

Stable marshes exist at an elevation that is supraoptimal for primary productivity. At this elevation, primary production is sensitive to changes in MSL because evapotranspiration and flood frequency interact to control the salt balance of the sediment (Morris, 1995). Adjustments in productivity and sedimentation rate help to maintain the marsh in a dynamic equilibrium with sea level, as long as the marsh elevation remains within a range that is supraoptimal for the vegetation. The equilibrium elevation varies inversely with the rate of sea-level rise. A loss in elevation can be counteracted by introducing species with greater tolerance to flooding. Soft engineering solutions to coastal flooding could incorporate planting of marsh vegetation in the intertidal zone for the purpose of promoting sedimentation and dissipating wave energy. A successful design would employ plant species that have varying degrees of tolerance to flooding, maximum drag, broad vertical ranges within the intertidal zone, and that form a successional series. In the context of management options for areas where there is significant risk of loss of cultural resources and valuable infrastructure, the use of introduced or transgenic species for coastline stabilization may be justified, provided that it can be demonstrated that the introduction would be ecologically benign and would not exact a greater cost from damage to biological resources. *Spartina* hybrids from San Francisco Bay resulting from hybridization of *S. alterniflora* and *S. foliosa* (Daehler & Strong, 1997) appear to have characteristics of a species well suited to compensate for sea-level rise.

Acknowledgements This work was supported by grants from the United States Geological Survey, the National Science Foundation, and the U.S. Environmental Protection Agency. The endorsement by these agencies of the conclusions of this research is neither implied nor intended. I thank Mark Brinson and two anonymous reviewers for providing helpful critiques.

References

Boumans, R. M. J. & J. W. Day Jr., 1993. High precision measurements of sediment elevation in shallow coastal areas using a sediment-erosion table. Estuaries 16: 375–380.

Bertness, M. D., 1991. Zonation of *Spartina patens* and *Spartina alterniflora* in a New England salt marsh. Ecology 72: 138–148.

Bertness, M. D. & A. M. Ellison, 1987. Determinants of pattern in a New England salt marsh plant community. Ecological Monographs 57: 129–147.

Bertness, M. D. & P. J. Ewanchuk, 2002. Latitudinal and climate-driven variation in the strength and nature of biological interactions in New England salt marshes. Oecologia 132: 392–401.

Cahoon, D. R., J. C. Lynch, P. Hensel, R. Boumans, B. C. Perez, B. Segura & J. W. Day Jr., 2002. A device for high precision measurement of wetland sediment elevation: I. Recent improvements to the sedimentation-erosion table. Journal of Sedimentary Research 72: 730–733.

Daehler, C. C. & D. R. Strong, 1997. Hybridization between introduced smooth cordgrass (*Spartina alterniflora*, Poaceae) and native California cordgrass (*Spartina foliosa*) in San Francisco Bay, California, USA. American Journal of Botany 85: 607–611.

Danielsen F., M. F. Sørensen, M. F. Olwig, V. Selvam, F. Parish, N. D. Burgess, T. Hiraishi, V. M. Karunagaran, M. S. Rasmussen, L. B. Hansen, A. Quarto & N. Suryadiputra, 2005. The Asian tsunami: A protective role for coastal vegetation. Science 310: 643.

Emery, N. C., P. J. Ewanchuk & M. D. Bertness, 2001. Competition and salt-marsh plant zonation: Stress tolerators may be dominant competitors. Ecology 82: 2471–2483.

Finley, R. J., 1975. Hydrodynamics and tidal deltas of North Inlet, South Carolina. In: Cronin E. (ed.), Estuarine Research, Vol. II. Academic Press, NY, 277–291.

Hutchinson, G. E., 1957. Concluding remarks. Cold Spring Harbor Symposia on Quantitative Biology 22: 415–442.

Krone, R. B. 1985. Simulation of marsh growth under rising sea levels. In Waldrop, W. R. (ed.), Hydraulics and Hydrology in the Small Computer Age. Hydraulics Div. ASCE.

Leonard, L. A. & M. E. Luther, 1995. Flow dynamics in tidal marsh canopies. Limnology and Oceanography 40: 1474–1484.

McKee, K. L. & W. H. Patrick Jr., 1988. The relationship of smooth cordgrass (*Spartina alterniflora*) to tidal datums: a review. Estuaries 11: 143–151.

Moller, I., T. Spencer, J. R. French, D. J. Leggett & M. Dixon, 1999. Wave transformation over salt marshes: A field and numerical modelling study from north Norfolk, England. Estuarine Coastal and Shelf Science 49: 411–426.

Morris, J. T., 1995. The salt and water balance of intertidal sediments: results from North Inlet, South Carolina. Estuaries 18: 556–567.

Morris, J. T., 2000. Effects of sea level anomalies on estuarine processes. In Hobbie, J. (ed.), Estuarine Science: A Synthetic Approach to Research and Practice. Island Press, 107–127.

Morris, J. T. in press. Estimating net pimary production of salt marsh macrophytes. In Fahey, T. J. & A. K. Knapp, (eds), Principles and Standards for Measuring Net Primary Production in Long-Term Ecological Studies. Oxford University Press, London.

Morris, J. T., D. Porter, M. Neet, P. A. Noble, L. Schmidt, L. A. Lapine & J. Jensen, 2005. Integrating LIDAR elevation data, multispectral imagery, and neural network modeling for marsh characterization. International Journal of Remote Sensing 26: 5221–5234.

Morris, J. T., P. V. Sundareshwar, C. T. Nietch, B. Kjerfve & D. R. Cahoon, 2002. Responses of coastal wetlands to rising sea level. Ecology 83: 2869–2877.

Pennings, S. C., M. B. Grant & M. D. Bertness, 2005. Plant zonation in low-latitude salt marshes: disentangling the roles of flooding, salinity and competition. Journal of Ecology 93: 159–167.

Redfield, A. C, 1972. Development of a New England salt marsh. Ecological Monographs 42: 201–237.

Stevenson J. C., L. G. Ward & M. S. Kearney, 1986. Vertical accretion in marshes with varying rates of sea level rise. In Wolfe, D. A. (ed.), Estuarine Variability. Academic Press, 241–259.

Temmerman, S., G. Govers, S. Wartel & P. Meire, 2003. Spatial and temporal factors controlling short-term sedimentation in a salt and freshwater tidal marsh, Scheldt estuary, Belgium, SW Netherlands. Earth Surface Processes and Landforms 28: 739–755.

Vogel, R. L., B. Kjerfve & L. R. Gardner, 1996. Inorganic sediment budget for the North Inlet salt marsh, South Carolina, U.S.A. Mangroves and Salt Marshes 1: 23–35.

MIX
Papier aus verantwortungsvollen Quellen
Paper from responsible sources
FSC® C105338

If you have any concerns about our products,
you can contact us on
ProductSafety@springernature.com

In case Publisher is established outside the EU,
the EU authorized representative is:
Springer Nature Customer Service Center GmbH
Europaplatz 3, 69115 Heidelberg, Germany

Printed by Libri Plureos GmbH
in Hamburg, Germany